Information Circular 9523

Guidelines for the Control and Monitoring of Methane Gas on Continuous Mining Operations

By Charles D. Taylor, J. Emery Chilton, Gerrit V.R. Goodman

DEPARTMENT OF HEALTH AND HUMAN SERVICES
Centers for Disease Control and Prevention
National Institute for Occupational Safety and Health
Office of Mine Safety and Health Research
Pittsburgh, PA • Spokane, WA

April 2010

This document is in the public domain and may be freely copied or reprinted.

Disclaimer

Mention of any company or product does not constitute endorsement by either the National Institute for Occupational Safety and Health (NIOSH) or the Mine Safety and Health Administration (MSHA). In addition, citations to Web sites external to NIOSH do not constitute NIOSH endorsement of the sponsoring organizations or their programs or products. Furthermore, NIOSH is not responsible for the content of these Web sites.

Ordering Information

To receive documents or other information about occupational safety and health topics, contact NIOSH at

>Telephone: **1–800–CDC–INFO** (1–800–232–4636)
>TTY: 1–888–232–6348
>e-mail: cdcinfo@cdc.gov
>
>or visit the NIOSH Web site at **www.cdc.gov/niosh**.

For a monthly update on news at NIOSH, subscribe to NIOSH *eNews* by visiting **www.cdc.gov/niosh/eNews**.

DHHS (NIOSH) Publication No. 2010–141

April 2010

SAFER • HEALTHIER • PEOPLE™

CONTENTS

GUIDELINES FOR THE CONTROL AND MONITORING OF METHANE GAS ON CONTINUOUS MINING OPERATIONS 1
   EXECUTIVE SUMMARY 1
   BACKGROUND 3
      USBM Research 1950 to 1969 4
      USBM Research 1970 to 1995 5
      NIOSH Research 1996 to Present 8
   TEST FACILITIES 8
      Introduction 8
      Ventilation Test Gallery 8
      Model Mining Machine 9
      Model Roof Bolter 10
      Methane Distribution in the Test Gallery 10
      Methane Monitors 11
      Airflow Monitors 12
      Data Acquisition and Analysis 12
      Summary 12
   MOVING AIR TO THE MINING FACE 13
      Introduction 13
      Maintaining Curtain/Tubing Setback Distance 13
      Extensible Systems—Jet Fans 14
      Effects of Airflow Velocity on Face Methane Levels 17
      Blowing Curtain and Tubing 18
      Estimating the Ventilation Flow Quantity Reaching the Face 19
      Airflow Between the Mouth of the Blowing Curtain and the Face 20
      Measuring Airflow at the Face 22
      Effect of Airflow on Methane Concentrations 23
      Summary 24
   EFFECT OF SCRUBBER OPERATION ON FACE AIRFLOW AND METHANE CONCENTRATIONS 24
      Introduction 24
      Scrubber Tests in Test Gallery 25

 Scrubber and Intake Flow, Balanced and Unbalanced Flows... 27
 Effect of Scrubber Flow on Intake Flow... 28
 Direction of Scrubber Exhaust, Effect on Intake Flow and Recirculation 30
 Scrubbers Used With Exhausting Face Ventilation 31
 Effect of Scrubber Use on Methane Levels Above the Machine....................................... 33
 Summary 35

EFFECT OF WATER SPRAYS ON FACE AIRFLOW AND METHANE CONCENTRATIONS 35
 Introduction.. 35
 Effect of Nozzle Direction 36
 Effect of Water Sprays on Intake Airflow 36
 Effect of Water Sprays on Airflow at the Face 36
 Effect of Sprays on Methane Distributions 38
 Summary 41

METHANE MONITORING 41
 Introduction.. 41
 Methanometers, Catalytic Heat of Combustion Sensors 41
 Methanometers, Infrared Sensors 42
 Methanometers, Use in Underground Coal Mines 42
 Measuring Methanometer Response Times... 43
 Methane Sampling Strategy, Sampling on the Mining Machine 49
 Summary 52

MEASURING GAS LEVELS OUTBY THE FACE 52
 Introduction.. 52
 Person-Wearable Methane Monitors 53
 Cap Lamp-Mounted Personal Monitor 55
 Summary 57

EVALUATING METHANE LEVELS IN AREAS OUTBY THE MINING FACE 57
 Introduction.. 57
 Underground Tests 58
 Summary 59

TECHNIQUES FOR MEASURING AIRFLOW 60
 Introduction.. 60
 Ultrasonic Anemometers 60
 Evaluating Performance of Ultrasonic Anemometers 61

    Low Air Velocity Measurements 61
    Comparing Instrument Airflow Readings in the Ventilation Test Gallery 62
    Comparison of One-, Two-, and Three-Axis Anemometers Behind the Blowing Curtain .. 63
    Instrument Orientation 63
    Flow Readings with One-, Two-, and Three- Axis Anemometers at the Mining Face 65
    Summary 67
  METHANE MONITORING DURING ROOF BOLTING 67
    Introduction 67
    Laboratory Tests 67
    Summary 69
  PRACTICAL GUIDELINES FOR THE CONTROL AND MONITORING OF METHANE GAS ON CONTINUOUS MINING OPERATIONS 70
  REFERENCES 71

FIGURES

Figure 1. Methane concentrations measured in face area [USBM 1958]. 4
Figure 2. Extensible curtain suspended from a rail and rollers. 6
Figure 3. Flooded bed scrubber design. 7
Figure 4. Spray fan system used to move air across the face. 7
Figure 5. Ventilation test gallery. 9
Figure 6. Model mining machine showing scrubber. 9
Figure 7. Water sprays on model mining machine. 10
Figure 8. Model roof bolting machine. 10
Figure 9. Gas manifold at the test gallery face. 11
Figure 10. Methane sampling system. 11
Figure 11. Overhead support system for anemometers. 12
Figure 12. Spiral reinforced tubing attached to mining machine. 14
Figure 13. Expansion of jet in mine opening. 14
Figure 14. Jet fan test area in ventilation test gallery. 15
Figure 15. Jet fan configurations. 15
Figure 16. Airflow profiles for three jet fan configurations. 16
Figure 17. Rigid ducting on jet fan used to reduce recirculation. 16
Figure 18. Use of curtain to reduce recirculation. 17
Figure 19. Face sampling locations for jet fan tests. 17
Figure 20. Effect of jet fan velocity on face concentrations. 18
Figure 21. Methane sampling locations for curtain/tubing tests. 19
Figure 22. Average face methane concentration with tubing or curtain. 19
Figure 23. Mining machine locations for measuring face airflow. 20
Figure 24. Sampling locations for 35-ft setback distances. 21
Figure 25. Flow profiles with 10,000 ft3/min curtain flow for 13- and 16½-ft entries at a 35-ft

curtain setback (A), 25-ft curtain setback (B), and 15-ft curtain setback (C). 22
Figure 26. Face velocities at a 25-ft setback for the 13-ft wide entry (A) and the 16½-ft wide entry (B). 23
Figure 27. Methane profiles with 10,000 ft3/min curtain flow for 13- and 16½-ft entries at a 35-ft curtain setback (A), 25-ft curtain setback (B), and 15-ft curtain setback (C) 24
Figure 28. Airflow measurement locations with mining machine at the face. 25
Figure 29. Airflow toward the face for 4,000-ft3/min intake (A) and 6,000-ft3/min intake (B). 26
Figure 30. Airflow parallel to face at location 1 (A) and location 2 (B). 26
Figure 31. Methane sampling locations. 27
Figure 32. Effects of intake and scrubber flow on methane concentrations. 28
Figure 33. Effect of scrubber flow on intake air quantity. 28
Figure 34. Test conditions while measuring flow at the regulator door and behind curtain. 29
Figure 35. Changes in intake flow measured at curtain and regulator due to operation of the scrubber. 29
Figure 36. Directions of scrubber exhaust. 30
Figure 37. Effect of scrubber exhaust direction on face concentration. 30
Figure 38. Airflows with blowing (A) and exhausting (B) conditions at a 35-ft setback distance. 31
Figure 39. Effects of scrubber use with blowing and exhausting face ventilation. 32
Figure 40. Effect of setback distance on methane concentration with scrubber operating. 32
Figure 41. Airflow patterns using a scrubber with exhaust ventilation (35- and 25-ft curtain setbacks). 33
Figure 42. Methane sampling locations above mining machine (scrubber tests). 33
Figure 43. Methane concentrations above mining machine. 34
Figure 44. Dilution of methane at intake flows of 4,000 ft3/min (A) and 6,000 ft3/min (B). 35
Figure 45. Sampling locations for water spray tests. 36
Figure 46. Airflow at mouth of blowing curtain with angled versus straight sprays and high versus low pressure. 37
Figure 47. Airflow with no water sprays operating at locations 1 and 2 shown in Figure 45. 37
Figure 48. Airflow at curtain (A) and off-curtain (B) side of face. 38
Figure 49. Methane sampling locations above mining machine (water spray tests). 39
Figure 50. Methane distribution, 4,000 ft3/min intake flow. 39
Figure 51. Methane distribution, 6,000 ft3/min intake flow. 40
Figure 52. Effect of intake flow (A), nozzle direction (B), and water pressure (C) on concentration versus distance from face. 40
Figure 53. Sampling head with catalytic heat of combustion sensor. 42
Figure 54. Components of infrared sensor. 42
Figure 55. Test box used for response time measurements. 44
Figure 56. Sensor heads for monitors A, C, and G. 45
Figure 57. Response times measured with calibration cup (A) and test box (B). 45
Figure 58. Sensor heads (A, C, and G) with and without the calibration cups attached. 46
Figure 59. Response times measured in test box (dust cap off). 47
Figure 60. Infrared sensor heads with dust caps. 47
Figure 61. Response times with dust caps on (left plot) and off (right plot). 47
Figure 62. Sampling locations on model mining machine. 48
Figure 63. Catalytic heat of combustion sensor head. 48

Figure 64. Data collected using three methanometers. 49
Figure 65. Methane sampling locations on mining machine and at face. 50
Figure 66. Comparison of measurements on mining machine. Locations given in Figure 65. ... 50
Figure 67. Comparison of face methane concentrations versus concentrations at location 4 (return air side, 5.5 ft from the face). 51
Figure 68. Concentrations measured at locations 4 and 6 (return air side, 5 and 6.5 ft from face, respectively). 52
Figure 69. Daily readings for person-wearable monitors. 53
Figure 70. Response times for person-wearable monitors. 54
Figure 71. Cap lamp-mounted methane monitor. 56
Figure 72. Instrument alarm times (calibration gas 0.6% to 2.5%). 56
Figure 73. Instrument readings for 10 workdays. 57
Figure 74. Methane sampling locations in one mine. 58
Figure 75. Accumulation of gas at two faces in one mine. 59
Figure 76. Ultrasonic sensor heads with flow components. 61
Figure 77. Low air velocity apparatus. 61
Figure 78. Airflow sampling locations behind the curtain. 62
Figure 79. Velocity readings for three flows measured with two, 3-axis anemometers. 63
Figure 80. Velocity readings for three different flows measured with 1-, 2-, and 3- axis anemometers. 63
Figure 81. Measuring a yaw angle. 64
Figure 82. Effect of yaw angle on anemometer readings. 64
Figure 83. Evaluating effect of tilt angle on airflow readings. 65
Figure 84. Effect of tilt angle on airflow readings. 65
Figure 85. Face airflow sampling locations. 66
Figure 86. Anemometer readings at face sampling locations. 66
Figure 87. Flow direction at face. 67
Figure 88. Methane sampling locations for roof bolter tests. 69
Figure 89. Effects of release location on concentration. 70

ACRONYMS AND ABBREVIATIONS USED IN THIS REPORT

CFR	Code of Federal Regulations
MSHA	Mine Safety and Health Administration
NIOSH	National Institute for Occupational Safety and Health
PRL	Pittsburgh Research Laboratory (NIOSH)

UNIT OF MEASURE ABBREVIATIONS USED IN THIS REPORT

cm^3	cubic centimeters
dB	decibels
dBA	decibels – 'A' weighted
ft^3/min	cubic feet per minute
ft/min	feet per minute
ft	feet
Hz	hertz
in	inch
mm	millimeter
pct	percent
psi	pounds per square inch
sec	seconds
u, v, w	flow vectors
µm	micrometer
θ	yaw and tilt angles

GUIDELINES FOR THE CONTROL AND MONITORING OF METHANE GAS ON CONTINUOUS MINING OPERATIONS

Taylor CD; Chilton JE; and Goodman GVR

National Institute for Occupational Safety and Health
Office of Mine Safety and Health Research
Pittsburgh, PA

EXECUTIVE SUMMARY

Until the early 1980s, mine face ventilation systems were designed for ventilating cutting depths up to 20 feet. Since that time, use of remotely operated mining machines have allowed cutting depths to increase to 40 ft, increasing concerns about the effects on methane levels at the mine face area. The principles for efficient methane control during deeper cutting remained the same, namely

- Move a sufficient quantity of intake air from the end of the tubing or curtain to the face.
- Mix intake air with methane gas liberated at the face.
- Move methane contaminated air away from the face.

However, when cutting to depths greater than 20 ft (known as deep-cut mining), airflow quantities reaching the face area often decreased because it was difficult to maintain tubing or brattice setback distances. Earlier research showed that use of machine-mounted scrubbers and water sprays increased airflow at the face area during deep cutting. NIOSH research examined how these and other factors affected face airflow. A full-scale ventilation test gallery was used to study how different operating conditions caused airflow patterns and methane distributions near the face to vary.

The research results showed that during deep-cut mining[1]

- Without additional controls, only a small percentage of the air delivered to the end of the tubing or curtain reached the face area.
- Operation of a machine-mounted scrubber increased airflow and reduced methane levels at the face area as long as the quantity of intake air delivered to the end of the curtain or tubing was not reduced.
- Operation of water sprays did not significantly increase the volume of air reaching the face but did improve mixing of methane and intake air at the face.

Methane monitoring requirements remained the same for deep cutting, but the possibility of rapidly changing conditions at the face increases the need for accurate estimates of face methane

---

[1] The findings and conclusions in this report are those of the authors and do not necessarily represent the views of the National Institute for Occupational Safety and Health. Mention of any company or product does not constitute endorsement by NIOSH.

concentration. Research examined currently available instrumentation and sampling methods for monitoring methane at the face.

The results from this NIOSH research program demonstrate how existing and new engineering controls can be used to reduce face methane levels. The sampling methods that were investigated can provide better ways to measure methane levels near the front of the continuous mining machine.

In this report several practical guidelines are recommended for controlling and monitoring methane levels in the face areas of underground coal mines. Most of the recommendations were based on studies conducted in the NIOSH ventilation test gallery.

- Free-standing fans can be used to ventilate empty headings in coal mines.
  - The fan nozzle should be designed to provide maximum throw distance.
  - Recirculation should be minimized by proper placement of fan inlet and or by placing curtains partway across the entry.

- With blowing systems, the single most important factor on face methane dilution is the velocity of the air directed toward the face.
  - For the same airflows, use of tubing rather than a curtain usually provides better control of face methane, especially at longer setback distances.

- With blowing and exhausting systems, and with the mining machine at the face, use of scrubbers increases the amount of intake airflow reaching the mining face.
  - Scrubber and spray systems should be designed to achieve efficient face ventilation for the effective removal of gas from the face.

- Measurement of airflow speed and direction between the curtain and the face helps to predict methane concentrations in the face area.
  - In empty entries, airflow velocity is much lower in narrower entries. More airflow should be provided during box cuts to prevent higher methane levels.

- Regardless of intake flow quantity, increasing scrubber flow will reduce face methane levels if recirculation is controlled. Recirculation can be controlled by
  - Minimizing leakage around the ventilation curtain.
  - Directing scrubber exhaust away from the blowing curtain. With exhaust systems the mouth of the curtain should always be outby the scrubber exhaust.

- Water sprays on the mining machine should be directed to provide the best airflow across the entire face.

- Methanometer response times can be measured using either of two techniques developed by NIOSH. Instruments with shorter response times more accurately measure current methane levels. Dust cap design has the greatest effect on response times.
  - When selecting a methanometer the dust cap design should be examined. The cap should protect the methane sensor from dust and water but not significantly increase the response time.

- Alternative methane sampling locations on the mining machine should be compared and selected based on the relative protection provided to the face workers.

- Mine personnel should be provided with methane monitors that can be worn while working in areas that cannot be regularly monitored. Audible, visual, and vibratory alarms for the monitors should be evaluated based on the environment in which the instruments are used.

- Miners must be safely removed from a mine without exposure to excessive methane following stoppage of a main fan.
    - Mines should be evaluated for the most likely area where methane gas can accumulate following stoppage of a main mine fan.

- In areas between the mouth of the ventilation curtain and the face, airflow direction is constantly changing and it is difficult to accurately measure flow velocity with a single-axis anemometer (e.g., a vane anemometer).
    - Following approval for underground use, multi-axis anemometers should be used to monitor airflow direction and velocity between the mouth of the ventilation curtain or tubing and the face. Multi-axis instruments should also be used to monitor flow at locations outby the mining face.

- During roof bolting, if it is not practical to monitor methane levels at the mining face, methane levels should be measured with a bolter machine-mounted monitor and a detector held 16 ft inby the last row of bolts using a extensible pole.

BACKGROUND

The U.S. Bureau of Mines (USBM) was formed in 1910 following a series of underground explosions that resulted in many fatalities and injuries [Kirk 1996]. Most of the explosions were the result of unsafe practices that occurred throughout the mines. Early USBM research focused on reducing the sources of the ignitions, such as non-permissible explosives and open cap lamps. As a result, there was a significant reduction in the number of explosions. However, it wasn't until years later that the importance of ventilation for controlling underground methane levels was recognized.

For many years, it was considered inevitable that explosive mixtures of methane would accumulate in the face areas of underground coal mines. Prior to mechanization, the rate at which methane was released into the mines was relatively slow. The amount of air provided to working places was small and was determined based on health requirements of the workers and animals and not for control of methane [USBM 1938; USBM 1940]. In fact, relatively small quantities of fresh air were needed to dilute the amount of gas released from the face.

The introduction of conventional mining methods was an important step in the mechanization of mining, which increased the rate of mining. But the intermittent nature of the conventional mining process usually allowed time for methane gas to be dispersed without the need for significant improvements in ventilation.

It was not until mines began to use continuous mining machines that higher methane levels became an issue and the need for better face area ventilation was recognized. The USBM recognized the importance of ventilation research.

> "... The need is especially great in connection with the introduction of the newly developed continuous mining machines in America which advances the working face so rapidly that very large volumes of methane may be released requiring special auxiliary ventilation in the working places to prevent occurrence of flammable concentrations of methane in the mine air" [USBM 1950].

The number of face ignitions increased as more continuous mining machines were placed underground. USBM sampling at mining faces showed the seriousness of the methane problem. Methane levels were dangerously high [USBM 1958]. In some cases (Figure 1), methane concentrations measured 20 ft from the mining face exceeded the lower explosive limit (5% by volume).

USBM Research 1950 to 1969

As early as the 1950s, there was general agreement that methane control on sections using continuous mining machines could only be achieved by providing more air to the face and improving mixing of the air at the front of the machine. Early research focused on ways to get more air to the mining face. This work showed the benefits of using blowing versus exhaust ventilation for methane control [Patterson 1961].

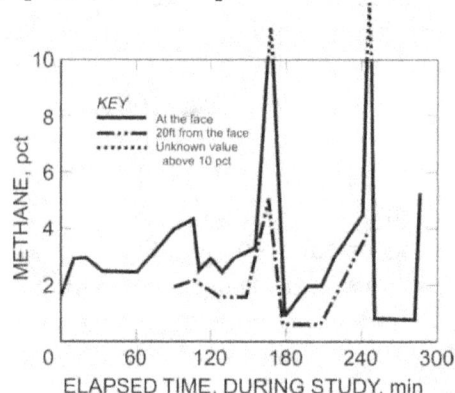

Figure 1. Methane concentrations measured in face area [USBM 1958].

Until the 1960s, curtains were the preferred method for directing air to the face. Most early curtain systems were also very inefficient, with as much as 95% of the air lost due to fabric porosity and leakage at the top and bottom of the curtain. Early research studies demonstrated how curtain installation techniques could be improved and recommended curtain materials that were less porous [USBM 1966].

The size of the mining machines made it difficult to build and maintain a curtain close to the face without interfering with machine movement in the entry. One solution was to use tubing, which required less space, to deliver air to the face. But ventilation tubing required use of auxiliary fans, which initially were banned from mines due to safety concerns.

The USBM obtained permission to use auxiliary fans in selected mines and demonstrated the advantages of using tubing for face ventilation. Tubing was easier to handle than curtain systems and required less space to install. In 1953, the Federal Mine Safety Code allowed the use of auxiliary fans. USBM testing examined the best ways to use tubing for face ventilation [USBM 1968].

Improving airflow to the face area was not enough. Some means for monitoring face methane levels were needed to assure methane levels were not dangerously high. The USBM recognized the need for better methane monitoring at the face area.

> "A desirable development would be the equipping of these [continuous] mining machines with a dependable automatic instrument for detection and warning of the presence of methane in concentrations approaching the lower explosive limit" [USBM 1950].

In 1958, the USBM initiated programs to develop a methane detecting system that would provide continuous monitoring of methane at the mining face [James 1959]. With this development, eventually all mining machines were required to be equipped with machine-mounted methane monitors [30 CFR 75.342(a)(1)].

USBM Research 1970 to 1995

The Federal Coal Mine Health and Safety Act of 1969 established operating standards for coal mine ventilation and set the maximum methane concentration permitted in the face area. Regulations published in Title 30, Part 75 of the Code of Federal Regulations mandated a minimum quantity of air reaching the face (3,000 ft^3/min), the maximum curtain setback distance (10 feet), and the maximum allowable methane concentration at the face (1% by volume).

The 1969 Act also set a maximum limit for airborne dust. Although the 1969 Act did not reduce the importance of ventilation for methane control, most mines in the 1970s placed an emphasis on improving ventilation for dust control. Several USBM studies emphasized the benefit of using exhaust ventilation for dust control [USBM 1969b; Mundell 1977]. But studies also showed that when exhaust ventilation was used and setback distances were greater than 10 ft, face area methane levels would increase [USBM 1969a]. Push-pull systems, which used both blowing and exhausting ventilation, were effective for both dust and methane control, but the additional required equipment made them impractical to use in most mining sections.

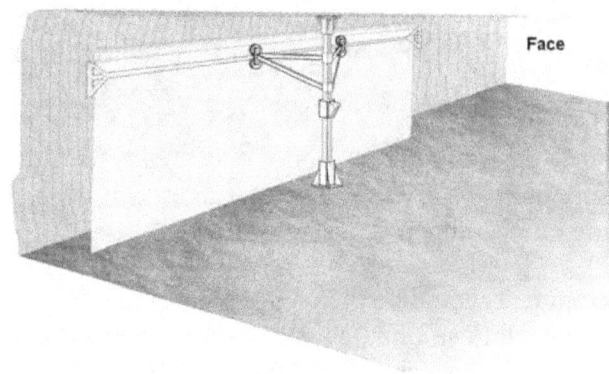

Figure 2. Extensible curtain suspended from a rail and rollers.

In most cases the curtain setback distance was set at 10 ft at the beginning of the cut but increased as the machine progressed into the cut. Tests were conducted to develop extensible ventilation devices that could be advanced during mining by workers from locations under supported roof systems. One example is the extensible curtain system illustrated in Figure 2. When using tubing, a 10- to 15-ft length of ventilation tube, known as a slider, is pulled from the outer slightly larger-diameter tube as mining advances.

Originally the extensible systems were designed for use with exhaust face ventilation. However, there was considerable leakage at the roof, making them much less effective for diluting methane at the face. An extensible curtain used with blowing ventilation was more efficient for maintaining delivery of airflow to the face area.

Although more airflow could be provided to the mining face area with blowing ventilation, compliance with the respirable dust standard was more difficult. In the early 1970s, the USBM and private industry began to experiment with machine-mounted dust collectors used with blowing face ventilation [Gillies 1982; USBM 1973]. These collectors

- Extracted dusty air from the face area through the use of fans.
- Removed the dust from the air with some type of filter or wetted fan.
- Exhausted the clean air at the rear of the mining machine.

Laboratory and underground studies showed that the flooded bed scrubber design in Figure 3 was best suited for dust reduction when mounted on a mining machine. When using the machine-mounted scrubber with blowing ventilation, it was possible to comply with the dust standard.

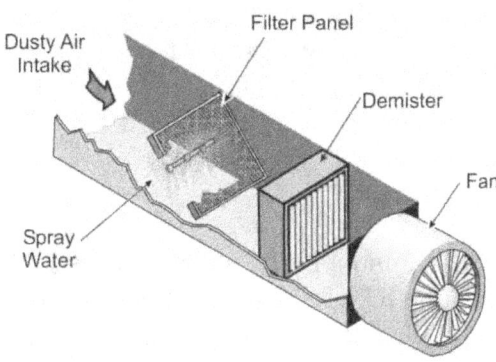

Figure 3. Flooded bed scrubber design.

On sections using exhaust ventilation, methane control was still a problem when setback distances exceeded 10 ft. Experiments had shown that machine-mounted water sprays used for dust control could move considerable quantities of air (Figure 4). When groups of these sprays were properly directed, airflow across the face area increased. Known as a spray fan system, these water sprays were effective for diluting and removing methane released at the face [Kissell 1979]. Most mines using exhaust ventilation installed some version of the spray fan system on their mining machines.

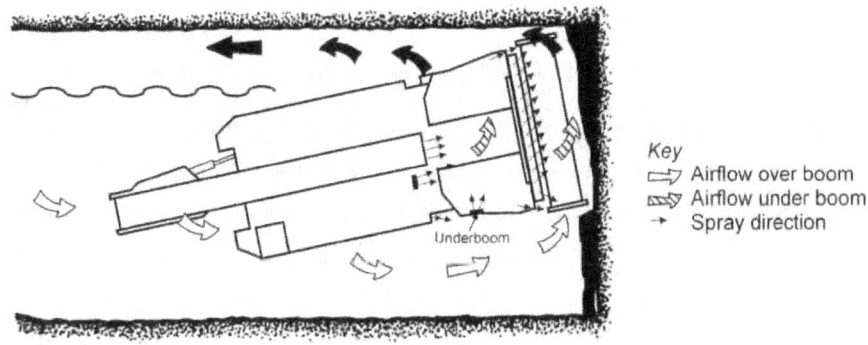

Figure 4. Spray fan system used to move air across the face.

By the 1980s, the USBM had demonstrated the basic engineering techniques needed to control both methane and dust at the mining face area. Typically, exhaust and blowing ventilation systems were used with 10- and 20-ft setback distances, respectively. Cutting depths had to be limited to 20 ft to prevent the miner operator from working under an unsupported roof.

During the 1980s, manufacturers began to equip continuous mining equipment with wireless remote control devices, which allowed the operator to be positioned outby and away from the machine. From this location, the operator could cut to depths exceeding 20 ft without being exposed to an unsupported roof. The Mine Safety and Health Administration (MSHA) permitted cutting depths greater than 20 ft, referred to as deep-cut mining, when the machine was operated remotely.

Most mines using deep-cut mining used blowing ventilation because it was easier to maintain adequate face area airflow at greater setback distances. Use of a machine-mounted scrubber further improved face area airflow. For mines using extended cutting with exhaust ventilation, use of the spray fan system improved face ventilation and reduced methane levels when ventilation setback distances exceeded 20 ft [Volkwein and Thimons 1986; USBM 1987].

NIOSH Research 1996 to Present

The Bureau of Mines continued its mining health and safety research program until 1996 when these functions were transferred to NIOSH. Since this time, research has evaluated how various engineering controls such as scrubber and water sprays can be used more effectively to improve face ventilation. Although the results were intended for deep-cut mining systems, the benefits can be realized by any operation using room and pillar mining systems. Methods for improving methane and airflow monitoring systems were developed and used to evaluate the effectiveness of mine face ventilation systems and improve the safety of underground workers.

TEST FACILITIES

Introduction

When possible, the effectiveness of a ventilation system to control methane should be evaluated in an underground mine. However, it is often preferable to conduct tests in a surface test facility where operating conditions can be controlled and varied as desired.

The NIOSH ventilation test gallery was initially designed as a facility where deep-cut face ventilation research could be conducted. In this test gallery, airflow conditions can be varied to simulate a wide range of ventilation conditions found in face areas. Methane is released at controlled rates and measured at multiple locations to study the effects of airflow on methane distributions inby the curtain. Airflow monitoring instruments have been adapted for measuring airflow direction and speed. The test gallery and monitoring equipment are described in the following sections.

Ventilation Test Gallery

The ventilation test gallery, a full-scale facility, is located in an "L-shaped" building (Figure 5). The height of the test area is 7 ft and the width is 16½ ft. Entry width can be reduced by building walls made of curtain and wood. An airtight wall that is built across the entry simulates the working face of a mine.

A vane-axial fan with a flow capacity of 12,500 ft^3/min draws air into and through the test gallery. Tests are conducted with either blowing or exhausting face ventilation using curtains or tubing to direct air to the simulated face. Airflow is varied by opening and closing regulator doors. An auxiliary fan with a vane-inlet controller is used to provide tubing airflow.

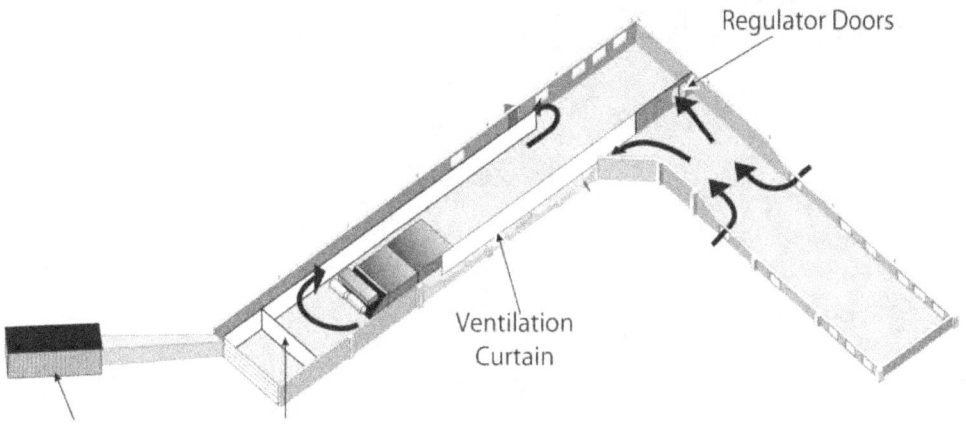

Figure 5. Ventilation test gallery.

Model Mining Machine

Full-scale model mining and roof bolting machines are used in the test gallery to simulate mining activity at the face. Both models are built of wood and covered with curtain material. The mining machine (Figure 6) has approximately the same outside dimensions as a Joy 14CM minus the rear loading boom. The miner was made in three wheel-mounted sections to facilitate movement.

The mining machine is equipped with a simulated dust scrubber and water sprays. The scrubber produces airflow similar to an actual scrubber. Two vane axial fans, with a combined flow capacity of 12,000 ft^3/min, draw air into inlets located near the front of the machine and exhaust it at the right rear of the machine. Orifice plates placed in the scrubber ducting are used to adjust flow quantity.

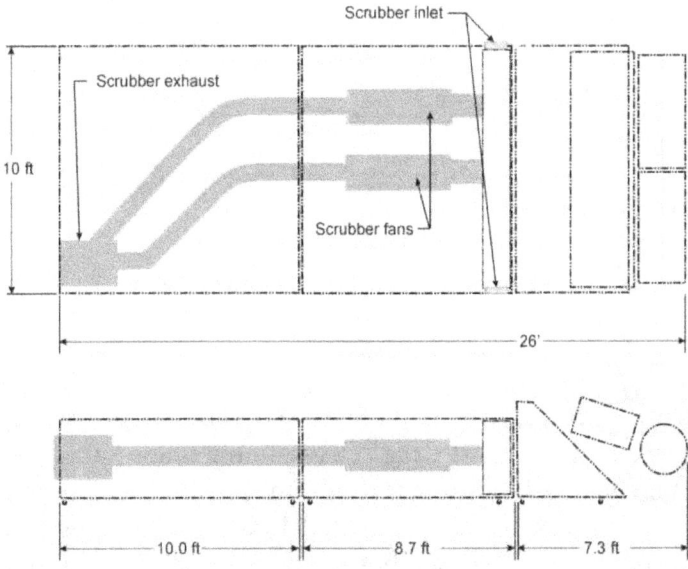

Figure 6. Model mining machine showing scrubber.

Spray manifolds are mounted on the top, sides, and under the boom of the mining machine (Figure 7). A centrifugal pump provides water pressures up to 170 psi.

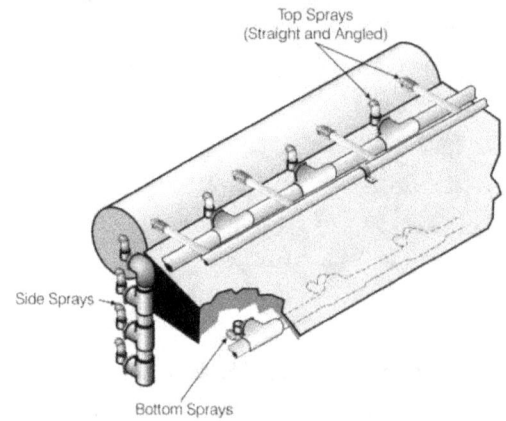

Figure 7. Water sprays on model mining machine.

Model Roof Bolter

The model roof bolting machine is slightly narrower than a typical bolting machine (Figure 8). Cutting booms and an automated temporary roof support extend from the front of the machine.

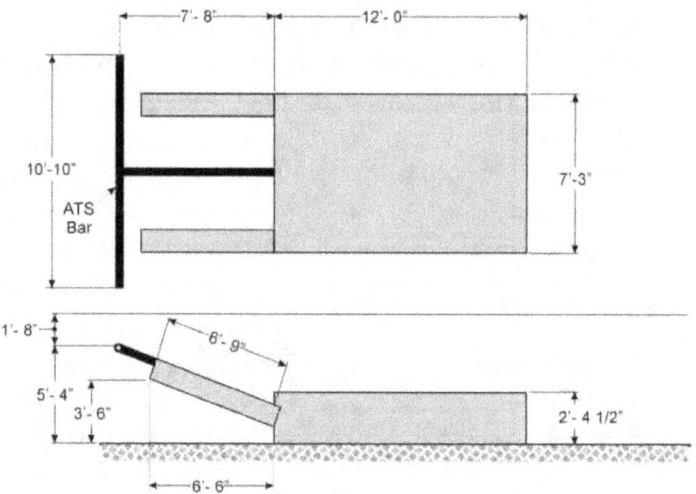

Figure 8. Model roof bolting machine.

Methane Distribution in the Test Gallery

Methane gas can be released from any location in an underground coal seam, but the location having the highest liberation rate is usually the active mining face. A uniform release of gas from the face is simulated in the test gallery using a manifold (Figure 9) consisting of four horizontally positioned 12-ft long copper pipes, which are located 4 inches from the simulated face. The pipes are equally spaced and perforated on top and bottom with 2-mm (1/16 in) diameter holes placed 2 in apart. To simulate a point gas source, methane is released through a hose nozzle. Gas flow rates are set and monitored using a rotameter. Unless otherwise noted, a constant methane flow was maintained for the duration of a set of tests. The methane gas is obtained from a commercial gas supplier.

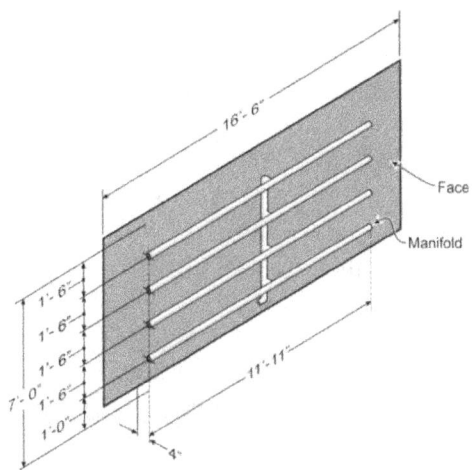

Figure 9. Gas manifold at the test gallery face.

Methane Monitors

Methane concentration can be simultaneously measured at up to 16 locations in the ventilation test gallery. A vacuum pump is used to draw air samples through plastic tubing from each sampling location in the test gallery to individual methane sensor heads (Figure 10). Data from the sensors are recorded by a computer-based data acquisition system.

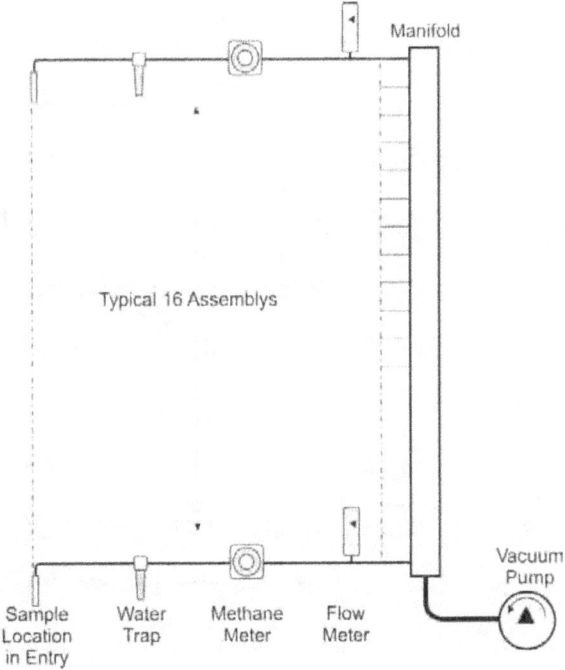

Figure 10. Methane sampling system.

Airflow Monitors

Ultrasonic anemometers are used for making most airflow readings in the ventilation test gallery. This instrument produces ultrasonic sound pulses which travel between transducers. The precise distances between the transducers are known. The measurement of airflow is based on the principle that the speed of a sound pressure wave varies with local air speed. Instrument calibration depends on accurate sensor spacing and transit time.

Normally, these anemometers are used to collect meteorological data from a single sampling location. Procedures were developed for using these instruments in the ventilation test gallery to measure both airflow direction and velocity [Taylor et al. 2003, 2002a, 2005; Hall and Timko 2005]. The data were used to evaluate the effect of airflow patterns on methane profiles near the mine face.

The sampling instruments are manually placed at sampling locations using either an individual support stand or suspended from an overhead support system (Figure 11). In either case the anemometer sensor head was positioned midway between the roof and floor. Software was developed to allow concurrent data acquisition from two anemometers.

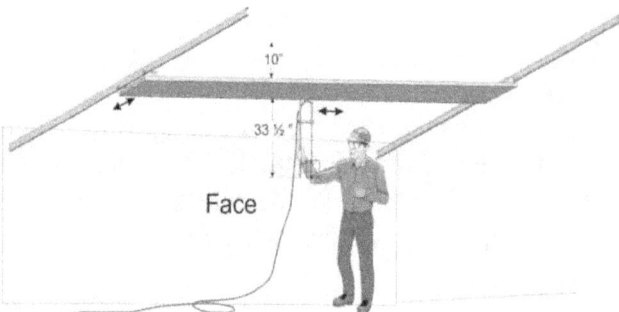

Figure 11. Overhead support system for anemometers.

Data Acquisition and Analysis

Data from methanometers and anemometers are recorded using computer-based data acquisition systems. Software used to collect and store data includes Labtech, Hyperterminal (Hilgraeve, Inc.), Windcom (Gill Instruments), and ANEMVENT 2003 (NIOSH-developed software). Most data are transferred to Microsoft Excel spreadsheets for analysis.

Summary

The NIOSH ventilation test gallery is a facility where techniques for methane control and monitoring are evaluated under a variety of conditions that simulate airflow near the working face of a continuous mining section. Airflow patterns and methane concentrations are studied in a detailed manner that is not possible in a working underground mine.

MOVING AIR TO THE MINING FACE

Introduction

Effective face ventilation requires that a sufficient quantity of intake air be delivered to the mining face in order to dilute liberated methane to a safe level. Federal regulations include the following requirements:

- Face ventilation control devices shall be used to provide ventilation to dilute, render harmless, and to carry away flammable, explosive, noxious, and harmful gases, dusts, smoke, and fumes [30 CFR 75.330].
- A minimum quantity of air (3,000 ft^3/min) is required at each face area [30 CFR 75.325].

A mine operation must specify in its ventilation plan the minimum quantity of air required to maintain methane levels below 1% at their working faces.

Early USBM research examined ways to deliver air to the end of the curtain or tubing with minimal losses. Recent NIOSH research has examined more effective ways to move air from the end of the curtain to the face. New monitoring instruments and sampling techniques made it possible to examine how operating conditions affect airflow inby the curtain or tubing.

Maintaining Curtain/Tubing Setback Distance

Regulations [30 CFR 75.330] require that curtain and tubing be installed no greater than 10 ft from the point of deepest face penetration unless another distance is approved in the ventilation plan. Most mines using blowing ventilation are granted permission to increase setback distance to 20 ft.

During deep cutting, it is more difficult to keep the end of the curtain or tubing close to the face without interfering with the mining operation or exposing workers to an unsupported roof. Earlier work examined how extensible curtain or tubing systems could be advanced manually to keep setback distances less than 10 ft. With these systems, leakage at the roof was a problem, they were labor intensive, and their use required periodic interruptions of the cutting sequence.

Several techniques for automatically advancing the ventilation system were proposed but never tested underground. One design deployed spiral reinforced tubing from the mining machine as it advanced into the coal face (Figure 12). A take-up device, similar to a shuttle car reel, could maintain constant tension on a cable that supported the tubing, causing it to retract as the machine backs away from the face. Another proposed technique that was tested in the laboratory used light-weight nylon tubing released from an auxiliary fan [Goodman et al. 1990]. Air pressure inflated the tubing as it moved away from the fan in the direction of the face. The tubing was manually pulled from the face when mining was completed.

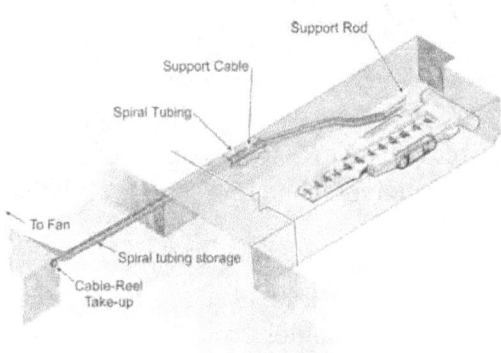

Figure 12. Spiral reinforced tubing attached to mining machine.

Extensible Systems—Jet Fans

A jet fan is a free-standing fan with no duct work and is usually placed on the upstream corner of the last open cross cut. It projects a high-velocity jet of air, which expands as it entrains the surrounding air into its stream. The jet expands until fresh air is flowing to the face in half the entry and returning as contaminated air through the other half (Figure 13).

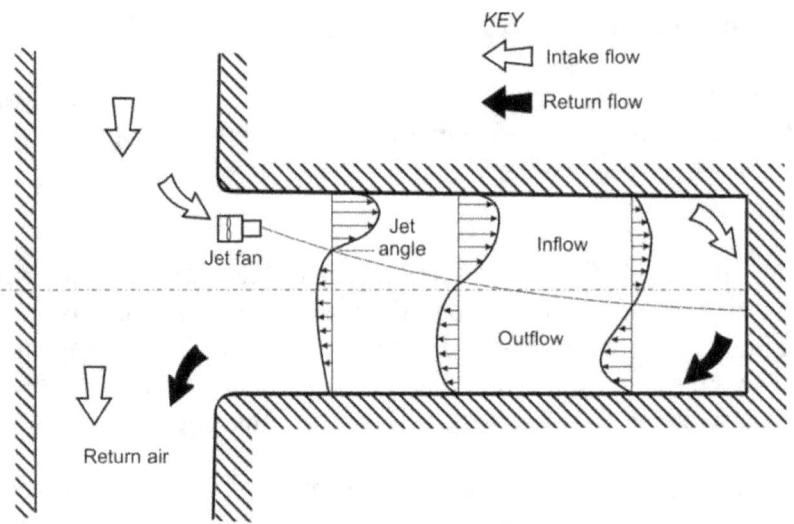

Figure 13. Expansion of jet in mine opening.

Underground tests have demonstrated the value of using jet fans for ventilating mines with large openings [USBM 1978; Engineers International, Inc. 1983]. Tests were conducted in the ventilation test gallery to determine if jet fans could be used to effectively ventilate mining entries having smaller cross-sectional areas, such as found in coal mines.

Tests were conducted in an area of the test gallery approximately 90 ft deep, 16½ ft wide and 7 ft high (Figure 14) [USBM 1992]. Exhaust airflow was 13,000 ft^3/min. An 18-in-diameter vane-axial fan with a maximum flow of 6,800 ft^3/min was used as the jet fan. The fan was positioned adjacent to the left (intake air) side of the entry, with the centerline of the fan exhaust 2 ft from the floor. The fan exhaust was directed straight toward the face, parallel to the sides of the entry.

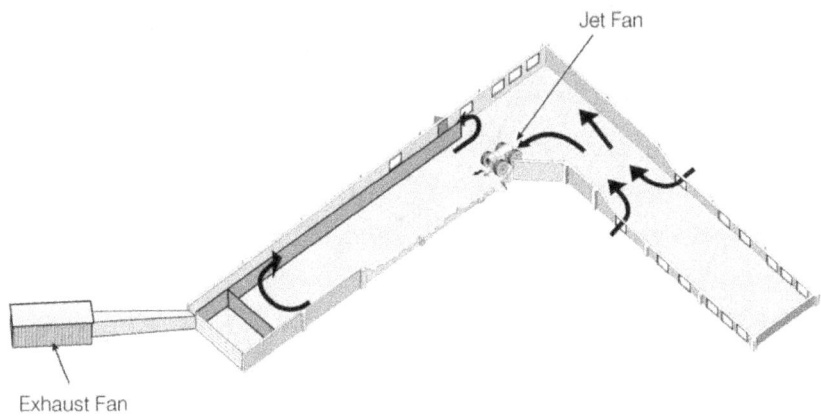

Figure 14. Jet fan test area in ventilation test gallery.

Three different jet fan configurations were tested (Figure 15). The nozzle was 16.5 in long with an outlet diameter that reduced from 18 in to 14.5 in. The flow extender had an 18-in diameter and was 48 in long. It had a set of internally mounted helical fins, which imparted a slight spinning action to the airflow.

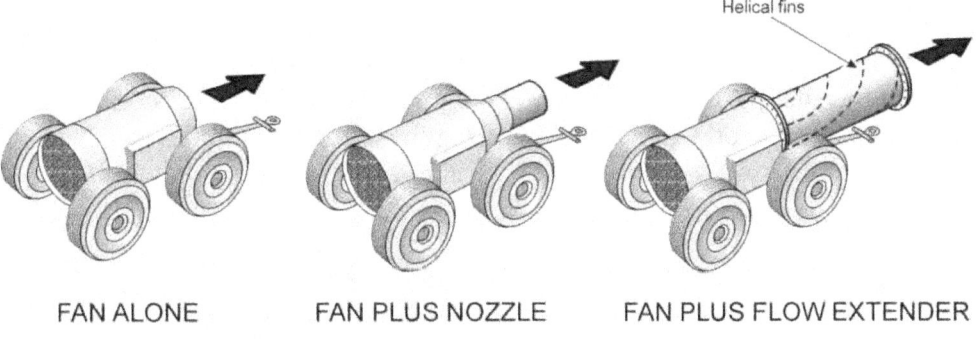

Figure 15. Jet fan configurations.

Using digital anemometers, velocity readings were taken at locations 10, 30, 50, and 70 ft from the fan outlet and at 2-ft intervals across the width of the test gallery. The data were used to determine the expansion angle and penetration depth of the air jet (Figure 16).

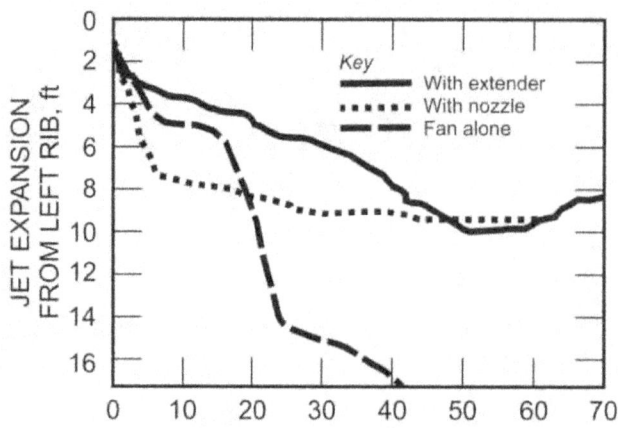

Figure 16. Airflow profiles for three jet fan configurations.

- With the fan alone, airflow was erratic, the expansion angle was approximately 26 degrees, and total penetration distance was less than 50 ft.
- The nozzle and flow extender increased the cohesion of the jet flow, resulting in higher velocities closer to the face. The angles ranged from 10 to 15 degrees, and the air jet flow penetrated to the face 70 ft away.

When using jet fans, recirculation is a concern. If the contaminated air from the face enters the fan, it can cause methane and dust concentrations at the face to increase [Campbell 1987]. Jet fan recirculation can be reduced or eliminated by placing the fan on the intake side of the last open cross cut and extending a short length of tubing from the fan exhaust to the mouth of the entry (Figure 17).

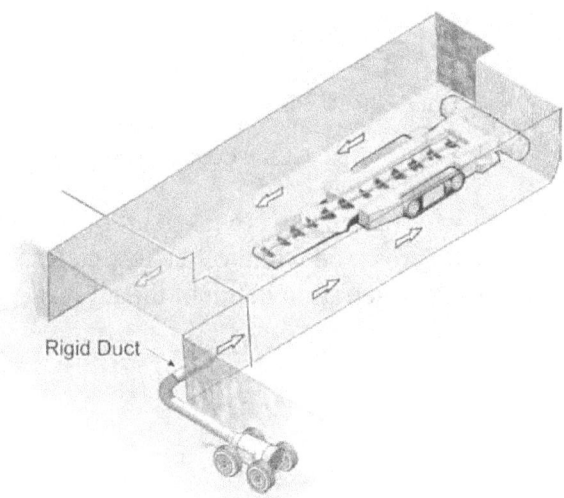

Figure 17. Rigid ducting on jet fan used to reduce recirculation.

A check curtain can also be used to direct air away from the fan inlet and reduce recirculation. In one series of tests, a check curtain was draped across the fan housing and was extended from the left rib to a point 8 ft across the heading. The effect of adding the curtain on fan recirculation is illustrated in Figure 18. In addition, the amount of air reaching the face doubled when the curtain was used.

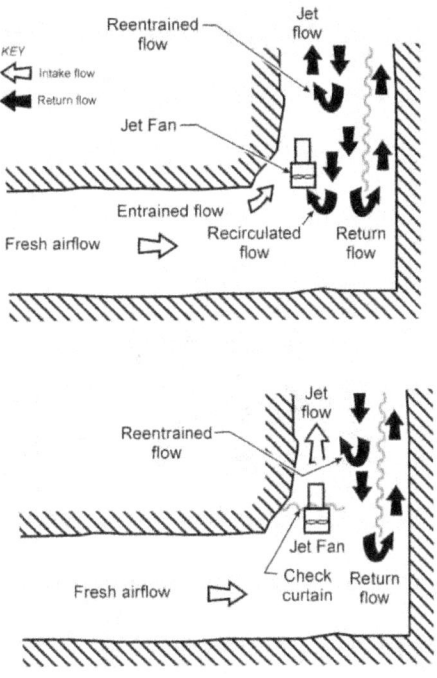

Figure 18. Use of curtain to reduce recirculation.

Effects of Airflow Velocity on Face Methane Levels

Tests were conducted to evaluate how face methane levels were affected by the air jet from the fan plus the extender [Taylor et al. 1992]. The fan with extender was located adjacent to the left (intake air) side of the galley and 40 ft from the face. Airflow velocity at the fan exhaust was varied from 2,040 to 3,850 ft/min using a vane inlet controller. Gas was released from the face manifold, and concentrations were measured at eight locations (Figure 19).

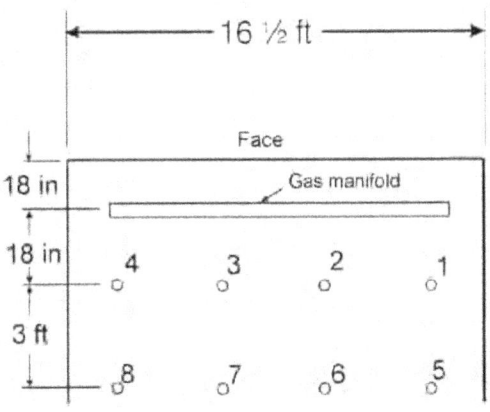

Figure 19. Face sampling locations for jet fan tests.

Methane concentrations increased at all sampling locations as airflow velocity decreased (Figure 20).

- For the eight sampling locations, the average concentrations increased when the velocity decreased from 3,850 to 2,040 ft/min.

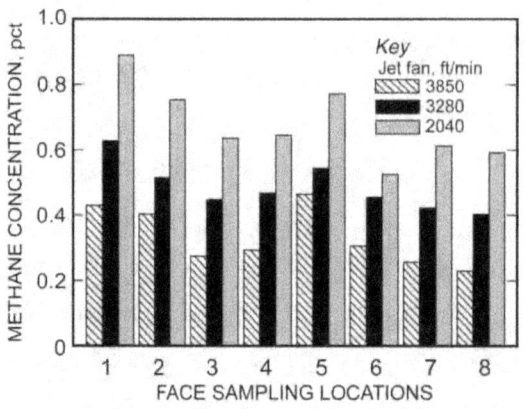

Figure 20. Effect of jet fan velocity on face concentrations.

Blowing Curtain and Tubing

Either curtain or tubing is used to direct intake air toward a mining face. The area behind a curtain is usually much larger than the cross sectional area of tubing. Therefore, for a given flow quantity, the velocity of air provided by tubing is much higher than that provided by a curtain. Tests were conducted to evaluate how this difference in velocities from curtain and tubing affects methane liberated from a mining face.

A series of tests were conducted with two methods of directing air to the face: 18-in diameter solid tubing and a 7-ft high curtain placed 2 ft from the side of the entry [Taylor et al. 1997]. Setback distances of 10, 30, and 50 ft with an intake flow of 5,500 ft^3/min were evaluated. Velocities at the mouth of the tubing and curtain were 3,125 and 393 ft/min, respectively.

The model mining machine was placed at the center of the face. Methane was released from the face manifold and measured at six locations near the face (Figure 21). The results are displayed in Figure 22.

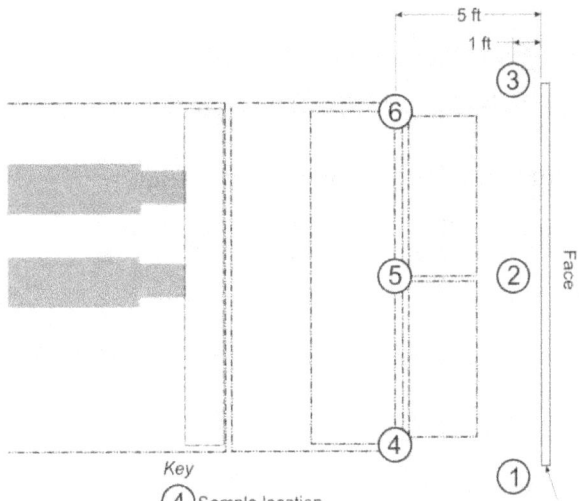

Figure 21. Methane sampling locations for curtain/tubing tests.

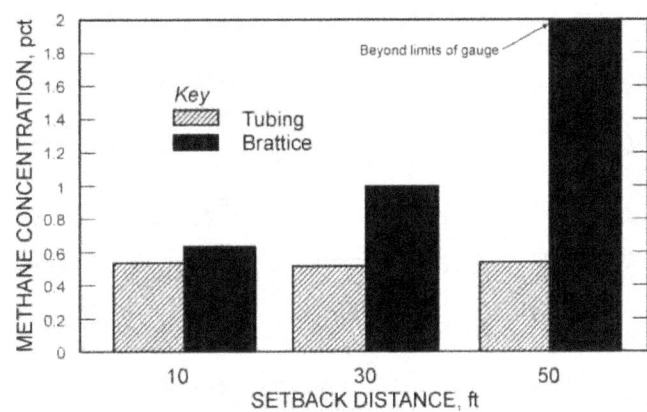

Figure 22. Average face methane concentration with tubing or curtain.

Methane levels increased as the curtain setback distance was increased from 10 to 50 ft. At the 50-ft setback, the test with curtain had to be terminated because methane concentrations at one or more of the sampling locations exceeded safe operating limits (2.5%) for the test gallery.

- Methane levels did not change significantly as tubing setback distance increased from 10 to 50 ft.
- The higher tubing velocity had a significant effect on lowering face methane levels.

Estimating the Ventilation Flow Quantity Reaching the Face

The quantity of intake air directed to a mining face is usually measured at the mouth of the ventilation tubing or curtain, but the quantity of intake air reaching the face cannot be precisely measured. The further the curtain or tubing mouth is from the face, the greater the uncertainty about the quantity of air reaching the face.

Tests were conducted in the ventilation test gallery [NIOSH 1999] to estimate the quantity of air delivered to a mining face when using blowing curtain and the following operating conditions:

- Machine at the end of box cut face (A in Figure 23).
- Machine at the beginning of slab cut (B in Figure 23).
- Blowing curtain with 50-ft setback.
- 4,000 or 10,000 ft^3/min measured at mouth of curtain.
- Scrubber off or at flow equal to intake flow.
- Spray on whenever scrubber was operating.

Methane gas was released from the face manifold at a known flow rate and measured at three locations (Figure 23) that were 1 ft from the roof and the face. The estimated flow (ft^3/min) reaching the face was calculated using measurements of methane flow rate through the manifold, the average face concentration, and the intake flow rate.

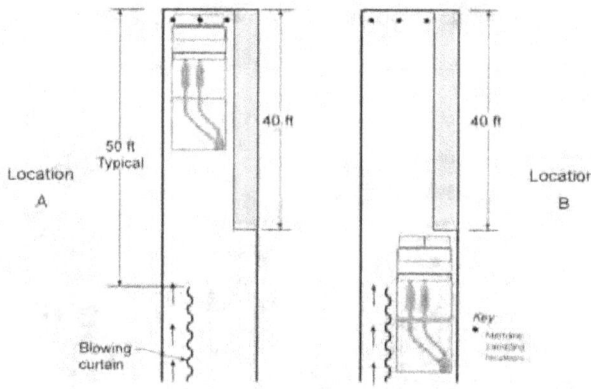

Figure 23. Mining machine locations for measuring face airflow.

The results indicated that use of a water spray and scrubber had a significant effect on airflow reaching the face only when the mining machine was located at the face (Figure 23).

- When the machine was at the end of the box cut (location A) and the scrubber and sprays were operating, more than 50% of the available air reached the face.

- When the machine was at the beginning of the slab cut (location B) and the scrubber and water sprays were operating, only 5% to 14% of the available air reached the face.

Airflow Between the Mouth of the Blowing Curtain and the Face

Past studies evaluated the movement of air between the end of the curtain or tubing by observing the movement of smoke generated with chemical tubes and making velocity measurements with vane anemometers or pitot tubes [USBM 1969a]. Smoke patterns were more difficult to evaluate where the airflow was turbulent. Accurate velocity readings were more difficult to obtain near the face because flow direction was constantly changing. Better instrumentation was needed for monitoring flow direction and velocity near the mining face.

One-, two-, and three-axis ultrasonic anemometers were obtained for making airflow measurements in the ventilation test gallery. The two- and three- axis instruments were used to measure flow speed and direction in the area between the face and the mouth of the curtain [Taylor et al. 2005].

The first tests were conducted in the test gallery where 16½- or 13-ft wide empty entries were simulated. Measurements were made with curtain setback distances of 15, 25, and 35 ft and intake flow quantities of either 6,000 or 10,000 ft^3/min. Most of the airflow sampling locations were 4 ft from one another and 2 ft from the sides and face of the entry.

Figure 24 shows the 36 sampling locations used for the 35-ft setback tests. Techniques for positioning the anemometers are described in Section 8 of this paper.

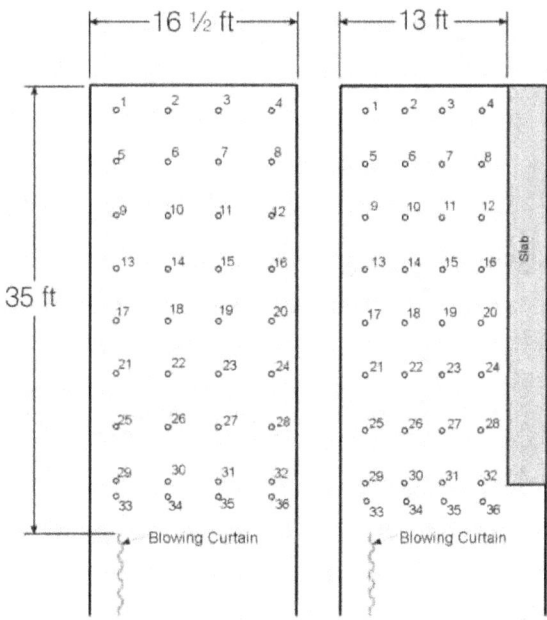

Figure 24. Sampling locations for 35-ft setback distances.

The individual measurements were used to draw the airflow profiles. Figure 25 shows airflow profiles for 10,000-ft^3/min intake flow tests. Vectors indicate each sampling location and direction of airflow. Vector length is proportional to the air velocity in the direction of the flow. Similar profiles were obtained for 6,000-ft^3/min intake flow tests, but the velocities were lower.

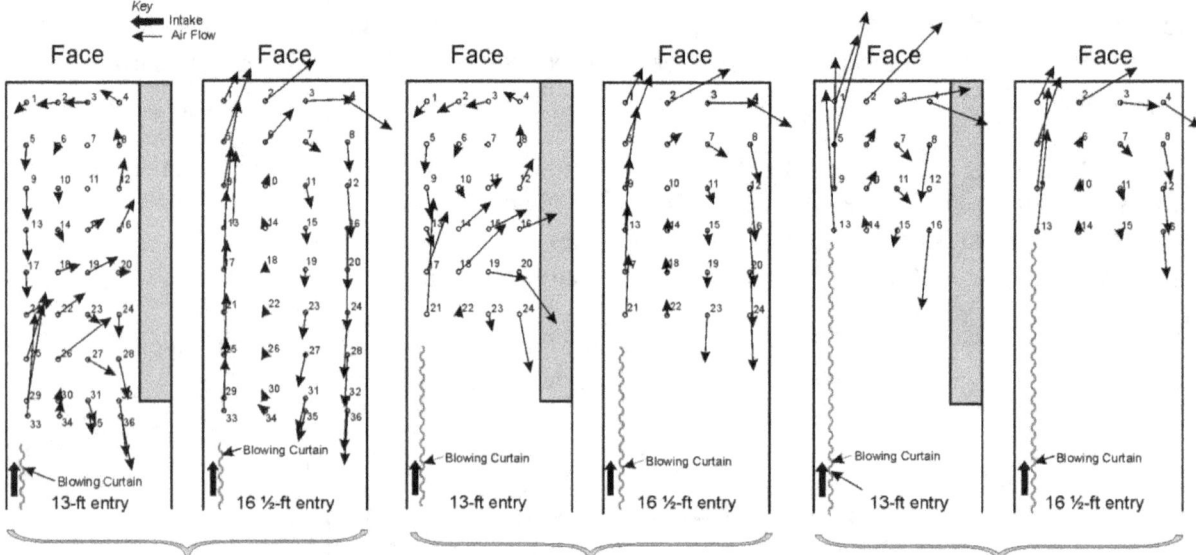

Figure 25. Flow profiles with 10,000 ft3/min curtain flow for 13- and 16½-ft entries at a 35-ft curtain setback (A), 25-ft curtain setback (B), and 15-ft curtain setback (C).

- Decreasing curtain setback distance increased air velocities.
- Entry width affected the flow patterns and velocities. For 25- and 35-ft setback distances
  - Flow patterns in the 13-ft wide entry resembled a figure "8" with flow moving right to left across the face.
  - Flow patterns in the 16½-ft wide entry resembled a "U" shape with air moving left to right across the face.
  - Air velocities were much higher in the 16½-ft wide entry.
- For the 15-ft setback distance, airflow moved left to right across the face for both entry widths.

Measuring Airflow at the Face

Methane concentrations are usually highest within 2 ft of the face. The effectiveness of a mine face ventilation system depends primarily on how quickly methane is diluted and removed from this area. Therefore, airflow velocities measured within 2 ft of the face (Figure 24, locations 1, 2, 3, and 4) were examined further. To simplify the comparison of the face velocities, the components of the flow moving parallel to the face were calculated for each of the sampling locations.

Figure 26 shows the air velocities measured at the four face locations when setback distance was 25 ft. Data are given for 6,000- and 10,000-ft^3/min intake air quantities.

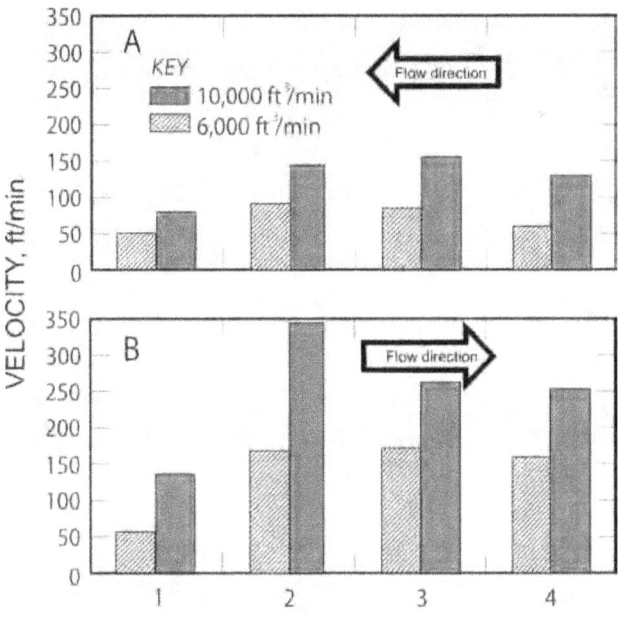

Figure 26. Face velocities at a 25-ft setback for the 13-ft wide entry (A) and the 16½-ft wide entry (B).

- Airflow velocities at the face increased with increasing intake flow.
- Airflow direction at the face was affected by entry width but unaffected by intake quantity.
- Airflow velocities at the face were much higher in the 16½-ft wide entry.

Effect of Airflow on Methane Concentrations

Methane and airflow readings were taken at the same locations with the same operating conditions in order to compare the effects of airflow on methane concentrations. Methane was released from the face manifold. Profiles were drawn (Figure 27) to show the distribution of methane between the mouth of the curtain and the face for a 10,000-ft^3/min intake flow. Similar profiles were obtained with the 6,000-ft^3/min intake flow, but the concentrations were higher.

The distribution of the methane between the curtain and the face and at the face locations 1 to 4 generally corresponded to the direction of the airflow.

- For left to right face airflow, the highest concentrations were generally on the right side of the entry.
- For right to left face airflow, the highest concentrations were generally on the left side of the entry.

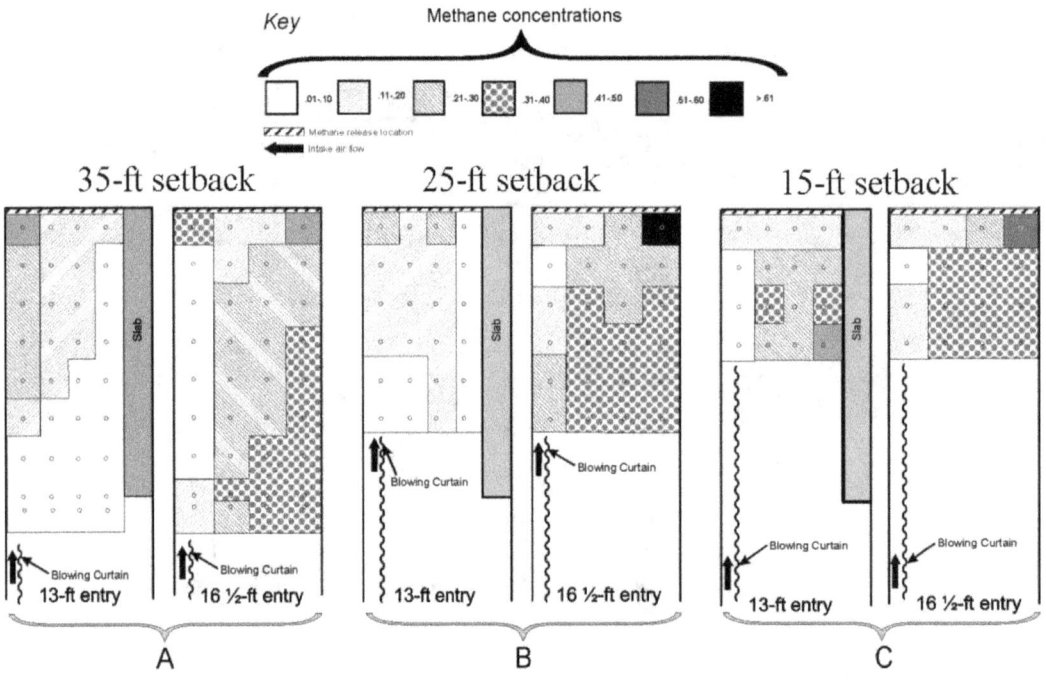

Figure 27. Methane profiles with 10,000 ft3/min curtain flow for 13- and 16½-ft entries at a 35-ft curtain setback (A), 25-ft curtain setback (B), and 15-ft curtain setback (C).

The methane concentrations (Figure 27) appeared to be similar for the 16½- and 13-ft wide entries. However, methane flow rates through the manifold had to be reduced 80% during tests in the 13-ft wide entry in order to maintain safe methane levels in the ventilation test gallery. If methane flows had been the same for all tests, concentrations would have been much higher in the 13-ft wide entry due to the lower face flow velocities.

Summary

Maintaining adequate intake airflow to the face is probably the most important factor affecting the control of methane gas liberated from the face and surrounding rock strata. Improving the effectiveness of a mine face ventilation system requires that either intake flow is increased or a greater percentage of the available intake air is delivered to the face. Increasing the intake air velocity at the mouth of the curtain or tubing improves the movement of air toward the face and increases the volume of air reaching the face. Decreasing entry width can have a major effect on flow patterns in the working entry by reducing air quantities reaching the face and reversing airflow direction across the face. Operation of machine-mounted water sprays and scrubbers can significantly increase the amount of airflow reaching the face.

EFFECT OF SCRUBBER OPERATION ON FACE AIRFLOW AND METHANE CONCENTRATIONS

Introduction

Scrubbers are used to remove dust from the air in the environment of the mining face. Dusty air from the face passes through and captured on a wetted filter and the cleaner air is exhausted at the rear of the mining machine. The scrubber moves a large quantity of air in the face area. Earlier work had shown that this air movement can improve the dilution and removal of methane gas from the face area.

Scrubber Tests in Test Gallery

The effects of scrubber operation on airflow near the mining machine were investigated in the ventilation test gallery [Taylor et al. 2006]. The model mining machine was located near the face and at the center of a 16½-ft wide entry. Airflow was monitored using three-axis ultrasonic anemometers placed at the six locations shown in Figure 28. Location 6 was at the mouth of the blowing curtain.

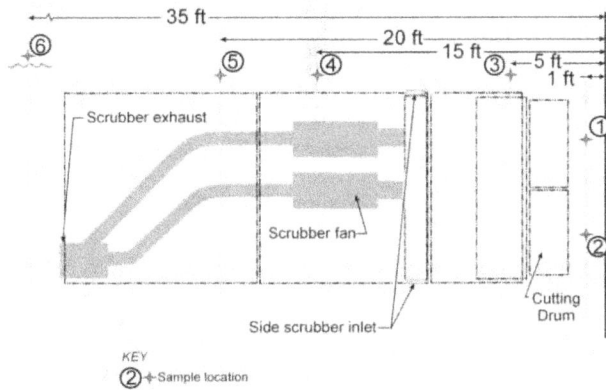

Figure 28. Airflow measurement locations with mining machine at the face.

Airflows were measured for different scrubber (0, 4,000, and 6,000 ft^3/min) and intake flows (4,000 and 6,000 ft^3/min). To simplify the comparisons, the flow components were calculated for either flow directed toward the face (locations 3 through 6) or parallel to the face (locations 1 and 2). Figure 29 shows flow velocities measured at locations 3 through 6 (5 to 35 ft from the face) for air moving toward the face.

With the scrubber off

- For both intake flows, velocities were nearly zero at distances of 5 and 15 ft from the face.
- At 5 ft, airflow reversed (negative velocity) and moved away from the face. This also occurred at 5 and 15 ft. in graph B.
- Figure-eight airflow pattern that occurs in the 13-ft wide entry is consistent with reversed flow at the left (intake air) side of the face.

With the scrubber on

- Airflow reached the face although the velocity decreased significantly 5 ft from the face.
- Airflow toward the face increased as scrubber flow increased (4,000 to 6,000 ft^3/min) but decreased at the 5 ft sampling location. Flow was reduced because air was diverted from the face to the scrubber inlet.

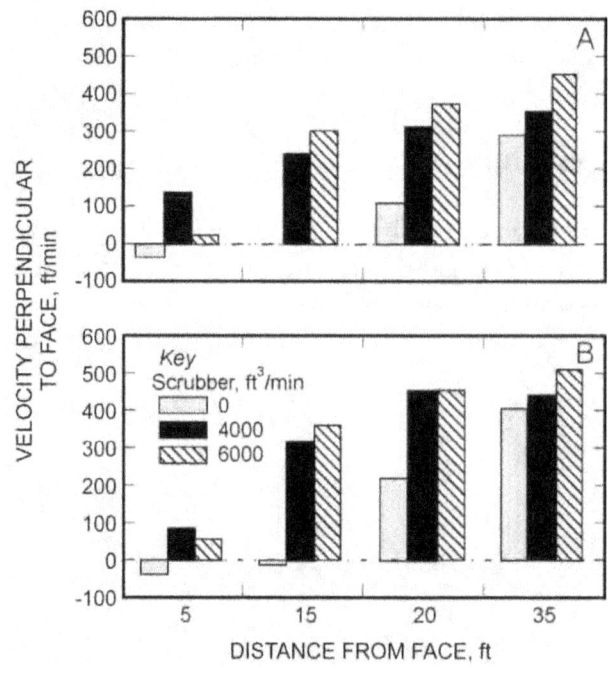

Figure 29. Airflow toward the face for 4,000-ft3/min intake (A) and 6,000-ft3/min intake (B).

Airflow moving parallel to the face, measured at locations 1 and 2 (left to right across the face), is shown in Figure 30. By convention, flow left to right is positive and right to left is negative.

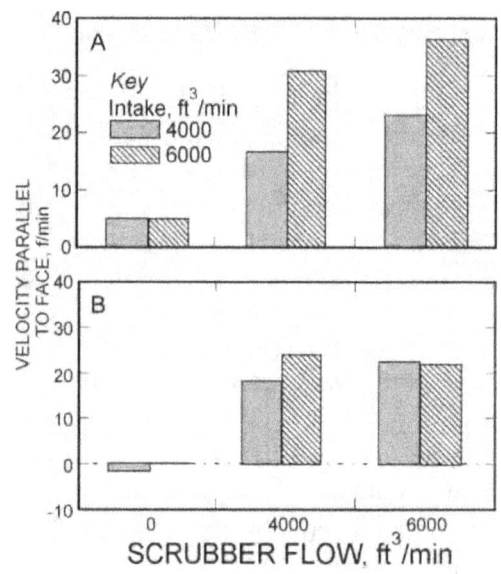

Figure 30. Airflow parallel to face at location 1 (A) and location 2 (B).

With the scrubber off, air velocity across the face was negligible (5 ft/min or less).
With the scrubber on, air moved across the entire face. In most cases, face flow increased with increasing intake flow.

Scrubber and Intake Flow, Balanced and Unbalanced Flows

Recirculation of air occurs at the mining face area when air flowing from the face is redirected by toward the face. Some recirculation is inevitable since it is not possible to separate airflow moving toward and away from the face. However, face ventilation procedures should minimize recirculation as much as possible to prevent contamination of intake air with methane-laden return air. Studies conducted in the test gallery evaluated how scrubber use affects recirculation and could potentially increase methane levels.

Earlier underground studies showed that methane levels did not increase during operation of a machine-mounted scrubber [Hadden and Hoover 1972]. However, concern about recirculation due to scrubber use continued. In many mines, it is common practice to maintain differences between scrubber and intake flow quantities of no greater than 1,000 ft^3/min. This practice is based on the assumption that when scrubber flow is greater than the intake flow, recirculation will increase resulting in higher methane levels.

Tests were conducted in the ventilation test gallery to compare face methane concentrations when scrubber and intake flows were balanced (equal) and unbalanced (unequal) [Taylor et al. 1997]. The effects on methane levels with scrubber flow exceeding intake flow were evaluated.

The model mining machine, equipped with water sprays and a dust scrubber system, was located at the center of the 16½-ft wide entry. Methane was released through the face manifold. The average methane concentrations measured at the six sampling locations shown in Figure 31 were used to evaluate scrubber operating conditions.

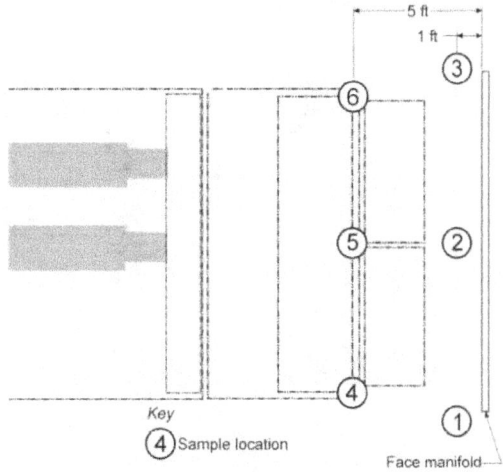

Figure 31. Methane sampling locations.

Measurements were taken for all combinations of intake and scrubber flows (6,000, 10,000, and 14,000 ft^3/min) and setback distances (25 and 35 ft). The average concentrations measured with the 25- and 35-ft setback distances were approximately equal and the data from the two setback distances were combined. The results are given in Figure 32.

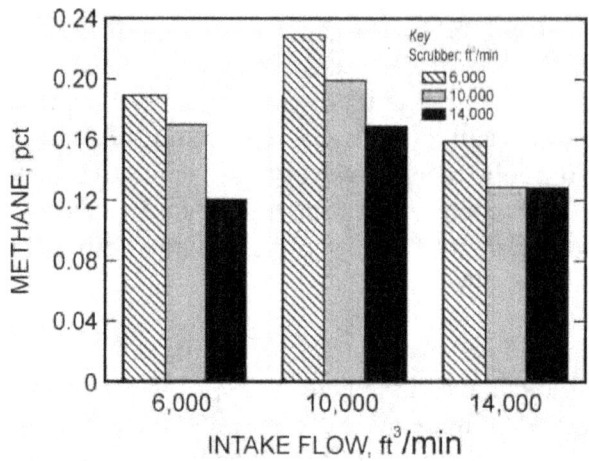

Figure 32. Effects of intake and scrubber flow on methane concentrations.

In most cases, methane levels decreased as scrubber flow increased. An unbalanced flow, with scrubber flow greater than intake flow, did not result in higher methane levels.

Effect of Scrubber Flow on Intake Flow

For each of the scrubber tests shown in Figure 33, intake airflow velocities were measured at the end of the blowing curtain, with and without the scrubber operating. Intake flows were always higher when the scrubber was operating, and intake flow increased as scrubber flow increased.

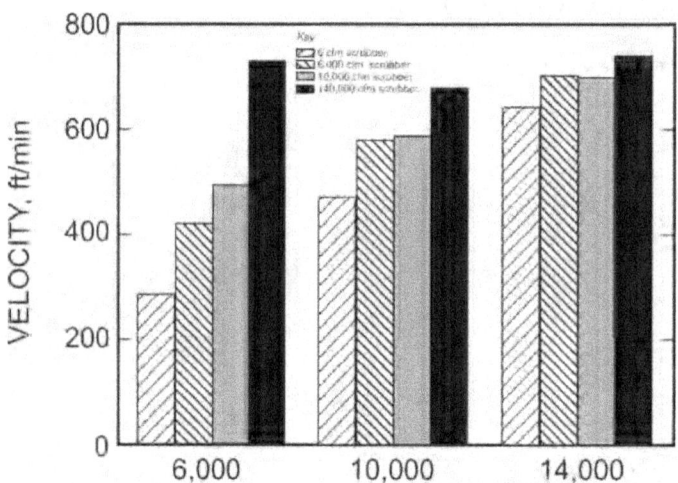

Figure 33. Effect of scrubber flow on intake air quantity.

Further tests were conducted to determine the source of the additional intake flow that resulted from scrubber operation [Taylor et al. 2006]. For these tests, one of the three regulator doors was opened completely (Figure 34) and the other two were closed completely. With the scrubber off, an outside door was opened enough to adjust intake flows behind the blowing curtain to either 4,000 or 6,000 ft^3/min. Intake and scrubber flow quantities were set at either (4,000 or 6,000 ft^3/min, while flow quantities were measured at the open regulator door and at the mouth of the blowing curtain.

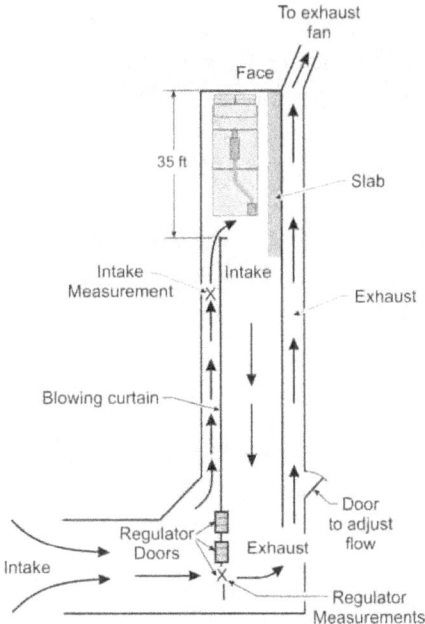

Figure 34. Test conditions while measuring flow at the regulator door and behind curtain.

Figure 35 shows changes in flow in the intake and at the regulator door for two test conditions:

- 4,000 intake [4K(I)] and 6,000 scrubber [6K(S)]
- 6,000 intake [6K(I)] and 4,000 scrubber [4K(S)]

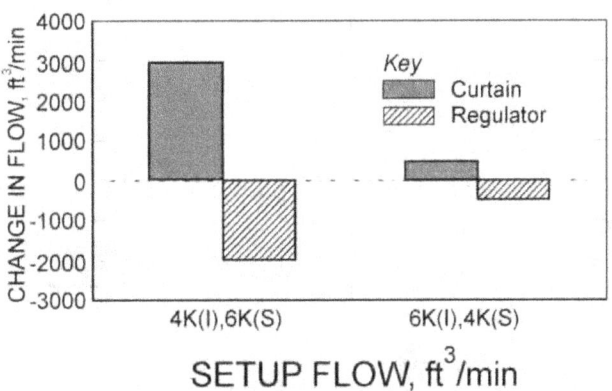

Figure 35. Changes in intake flow measured at curtain and regulator due to operation of the scrubber.

During scrubber operation, intake flow increased and flow through the regulator decreased.

- Intake flow increased more when scrubber flow was greater than the intake flow.
- The increase in intake flow was slightly greater than the decrease in flow through the regulator door.
    - Increased curtain flow was primarily due to air removed from the regulator.
    - Additional flow behind the curtain was due to leakage of air around the curtain.
    - Although scrubber use increased flow of uncontaminated air at the mouth of the

blowing curtain, leakage around the curtain added contaminated air to the intake air. Too much leakage could result in higher face methane levels. Maintaining intake flows higher than scrubber flows will reduce methane leakage around the curtain.

Direction of Scrubber Exhaust, Effect on Intake Flow and Recirculation

Most air from the scrubber exhaust moves straight back through the entry toward the return, but some of this air can recirculate back toward the face. The amount of recirculation depends on several factors including the direction and location of the scrubber exhaust relative to the mouth of the ventilation curtain. Tests examined the effects of the direction of the scrubber exhaust air on recirculation.

The blowing curtain setback distance was 35 ft, and intake flow quantity was 7,000 ft^3/min. A louver panel placed over the scrubber exhaust allowed the direction of the scrubber exhaust to be varied in the vertical and horizontal directions (Figure 36).

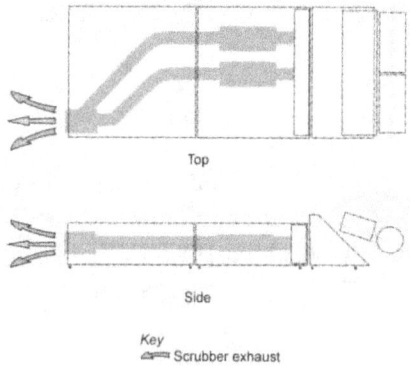

Figure 36. Directions of scrubber exhaust.

Methane gas was released from the face manifold. The average concentration was determined for six locations near the face as shown in Figure 31. Figure 37 shows the average concentrations for each of the exhaust directions.

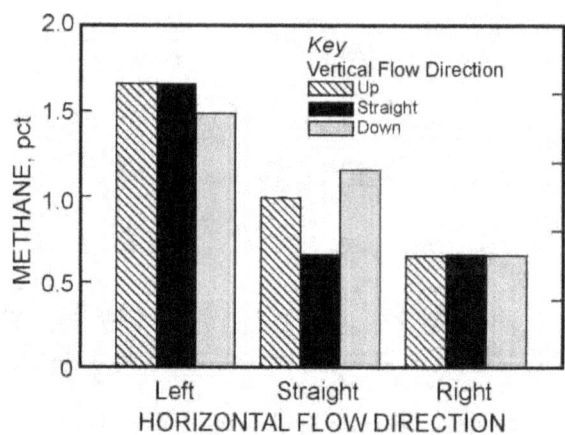

Figure 37. Effect of scrubber exhaust direction on face concentration.

- Methane concentrations were highest when the scrubber exhaust was directed toward the left (intake) side of the entry.
- Varying the vertical direction of the scrubber exhaust had little effect on methane concentrations.
- To reduce face methane concentrations
  - Do not direct scrubber exhaust toward the intake side of the entry. Directing flow toward the intake side of the entry interferes with and reduces flow moving toward the face.
  - Whenever possible, the scrubber exhaust and intake curtain should be placed on opposite sides of the entry.

Scrubbers Used With Exhausting Face Ventilation

Although machine-mounted scrubbers were initially designed to be used with blowing ventilation, mining sections with exhaust ventilation now also use them. Tests were conducted in the ventilation test gallery to compare the effectiveness of scrubbers used with blowing and exhausting mine face ventilation systems [Taylor et al. 1996].

Similar blowing and exhausting mine face ventilation systems were set up in the ventilation test gallery (Figure 38). Methane concentrations were measured for the same set of operating conditions. The blowing and exhausting curtains were constructed 3 ft from the left and right sides of the entry, respectively, and the mining machine was positioned at the center of a 16½-ft wide entry. Tests were conducted with setback distances of 35 and 25 ft and intake flows of 6,000, 10,000, and 14,000 ft^3/min. Scrubber and intake flows were equal.

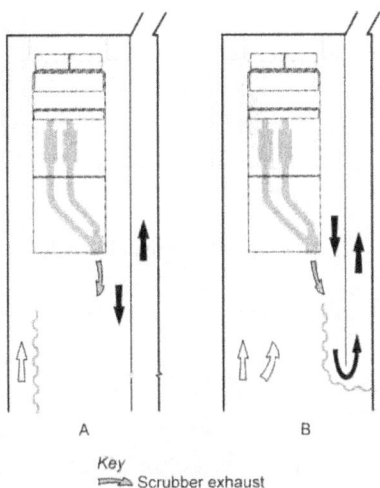

Figure 38. Airflows with blowing (A) and exhausting (B) conditions at a 35-ft setback distance.

Methane gas was released from the face manifold and gas measurements were made at the six locations near the face (Figure 31). Figure 39 shows the average concentrations for blowing and exhausting face ventilation.

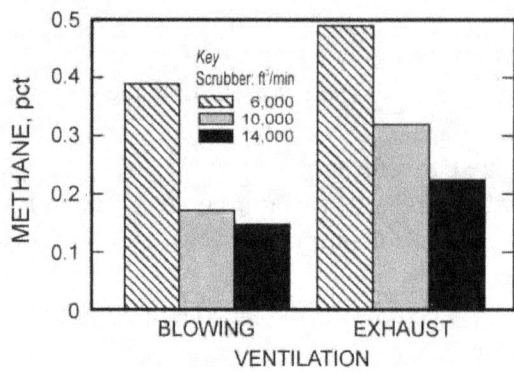

Figure 39. Effects of scrubber use with blowing and exhausting face ventilation.

- For the same operating conditions, face methane levels were lower when using blowing ventilation than when using the exhausting system.
- Increasing scrubber flow resulted in lower methane concentrations for both blowing and exhausting ventilation systems.

However, reducing curtain setback distance decreased face methane concentrations for the blowing system and slightly increased concentrations for the exhausting system (Figure 40).

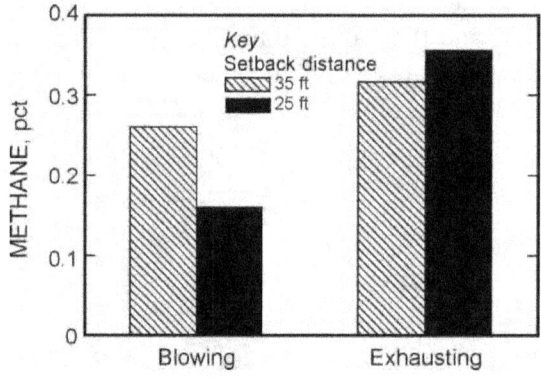

Figure 40. Effect of setback distance on methane concentration with scrubber operating.

- Reducing curtain setback distance from 35 to 25 ft decreased face methane levels for blowing ventilation.
- Reducing curtain setback distance from 35 to 25 ft slightly increased face concentration with exhausting face ventilation. At the 25-ft setback distance, the scrubber exhaust *was outby* the mouth of the curtain. Methane concentration increased because
  o Some of the scrubber exhaust re-circulated back to the face.
  o Airflow from the scrubber interfered with movement of intake air toward the face.

Figure 41 illustrates how the airflow patterns near the scrubber exhaust changed for the 25- and 35-ft curtain setbacks.

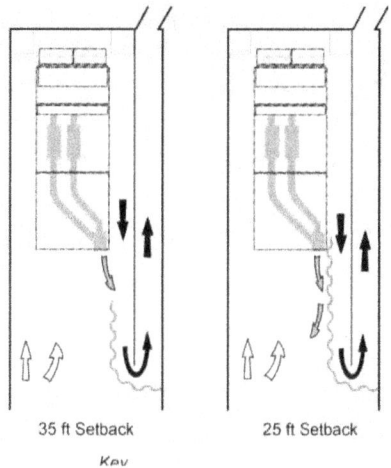

Figure 41. Airflow patterns using a scrubber with exhaust ventilation (35- and 25-ft curtain setbacks).

Effect of Scrubber Use on Methane Levels Above the Machine

Earlier work examined the effects of face airflow patterns on methane distributions in an empty entry. Tests were conducted to determine how scrubber operation affects methane levels above the mining machine, in the area between the face and blowing curtain [Taylor et al. 2006]. Methane levels were measured at 21 locations above the mining machine (Figure 42), which was located at the center of the 13-ft wide entry. The sampling locations were 4 to 4½ ft from one another and 2 ft from the sides and face of the entry. Curtain setback distance was 35 ft. Intake flows were 4,000 and 6,000 ft^3/min. The scrubber was either off or operated at flows of 4,000 or 6,000 ft^3/min.

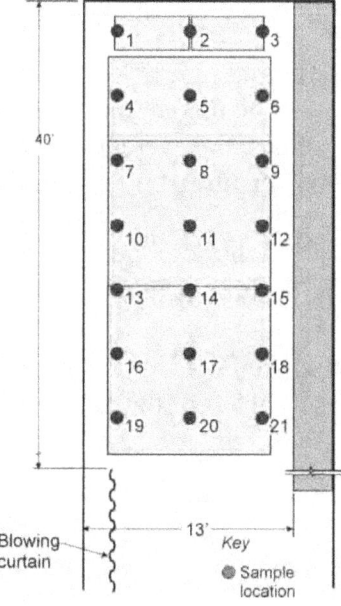

Figure 42. Methane sampling locations above mining machine (scrubber tests).

Methane was released from the face manifold. The distributions of the methane over the mining machine are shown in Figure 43 for intake flows of 4,000 and 6,000 ft³/min.

- Increasing scrubber flow decreased methane concentrations at most locations above the machine.
- Increasing intake flow decreased methane concentrations at most locations above the machine.
- The highest methane concentrations generally occurred in the corner of the face opposite the blowing curtain.

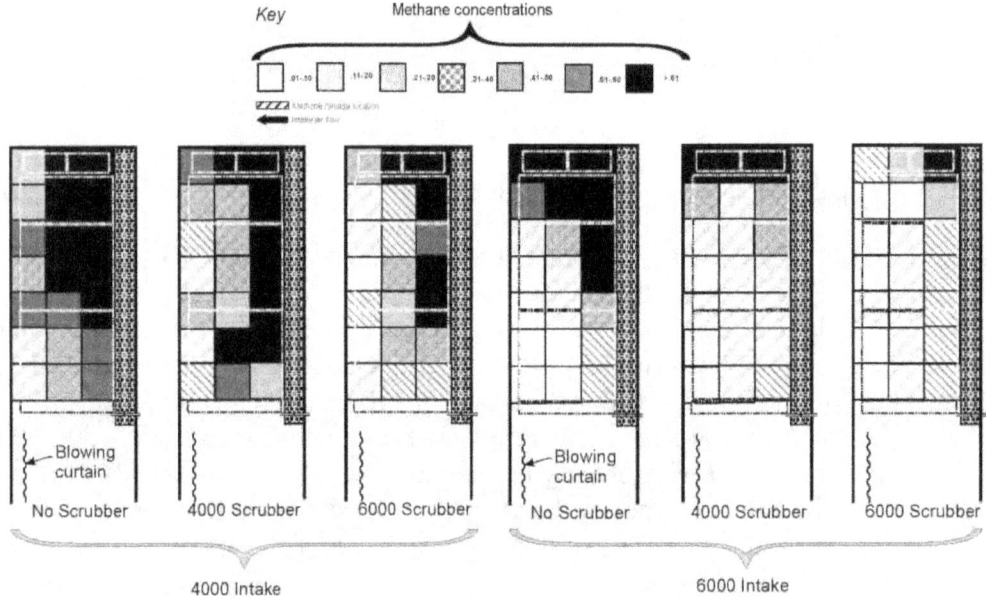

Figure 43. Methane concentrations above mining machine.

As methane gas moves further from a mining face it is diluted by the intake air. At some distance from the face, intake air and methane will be mixed uniformly and no further dilution will occur. To illustrate this progressive dilution of gas, the average concentration for each horizontal row of three samples was averaged and plotted versus distance from the face (Figure 44).

- With 4,000 ft³/min intake flow and the scrubber off, gas dilution takes place slowly.
- With 6,000 ft³/min intake flow, dilution takes place more quickly with and without the scrubber.
- With 4,000 and 6,000 ft³/min intake flow, scrubber use increased the rate of methane dilution, especially within the first 15 ft of the face.

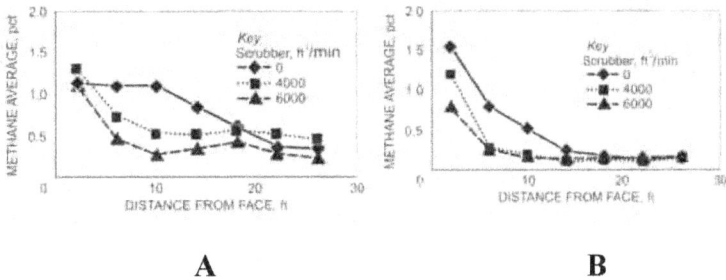

Figure 44. Dilution of methane at intake flows of 4,000 ft3/min (A) and 6,000 ft3/min (B).

Summary

Scrubbers can create significant air movement that can improve the dilution and removal of methane gas from the face area.

- Operation of machine-mounted dust scrubbers helps reduce face methane levels by
    - Increasing the volume of air reaching the face.
    - Improving the flow of air across the face.
    - Preventing formation of a figure-eight airflow pattern that results in lower air velocities at the face.
- Scrubber use increases the potential for recirculation of air to the face, but these effects are normally offset by higher air velocities and greater intake flow to the face.
- The potential for recirculation is greater if the scrubber exhaust interferes with intake air moving toward the face (exhaust and blowing systems).
- Scrubber use can improve face ventilation for exhaust and blowing mine face ventilation systems.
- Scrubber use reduces methane levels above the machine and at the face.

Note: These tests did not consider the effects of changes in scrubber and intake flows on airborne dust levels and, therefore, should not be used alone to develop a ventilation system for dust control.

EFFECT OF WATER SPRAYS ON FACE AIRFLOW AND METHANE CONCENTRATIONS

Introduction

Machine-mounted water spray systems are used primarily for dust control. The water delivered through the spray nozzles wets the coal and helps prevent suspension of dust. However, Kissell [1979] demonstrated that water sprays act as small fans and move air. This airflow helps dilute and remove methane from the face area. Water sprays can be grouped to direct airflow across the mining face. These "spray fan systems" are now installed on many mining machines. Research conducted in the ventilation test gallery examined how sprays installed on the mining machine affect airflow patterns and methane distributions in the face area [Chilton et al. 2006].

Effect of Nozzle Direction

Machine-mounted water sprays were mounted on the mining machine, which was located at the center of a 13-ft wide entry. The blowing curtain setback distance was 35 ft, and the intake flows were either 4,000 or 6,000 ft^3/min.

For one series of tests, only the two top-mounted spray manifolds with hollow cone nozzles were used. The nozzles in one manifold were directed straight toward the face. In the second manifold, the ten nozzles were directed 30 degrees to the right (Figure 7), toward the return side of the entry. Tests were conducted using high (174 psi) and low (70 psi) water pressures.

Ultrasonic anemometers were used to monitor airflow at the three locations shown in Figure 45. Location 3 was in the mouth of the blowing curtain.

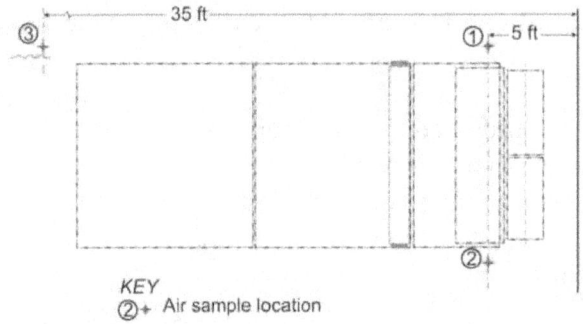

Figure 45. Sampling locations for water spray tests.

Effect of Water Sprays on Intake Airflow

Airflow velocities were measured at the mouth of the blowing curtain with the sprays turned off and with both spray configurations (straight and angled) operating at high and low pressures (Figure 46).

- Water sprays had no effect on intake flow measured at the mouth of the blowing curtain.

Effect of Water Sprays on Airflow at the Face

Air velocities were measured concurrently at two locations at the face—the curtain side (location 1) and the off-curtain side (location 2) (Figure 45). To simplify the comparisons, the component of the velocity moving perpendicular to the face was calculated for each test. By convention, an airflow moving away from the face was given a negative velocity. Figure 47 shows velocities measured with no sprays operating.

- With no sprays operating, flow was toward the face on the off-curtain side of the face (location 2) and away from the face on the curtain side of the face (location 1).
- Flow moved from right to left across the face and increasing intake flow increased air velocities moving toward and away from the face but did not affect the flow pattern.

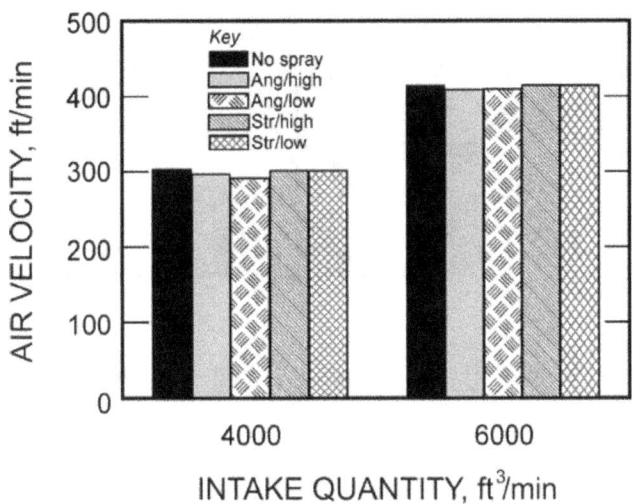

Figure 46. Airflow at mouth of blowing curtain with angled versus straight sprays and high versus low pressure.

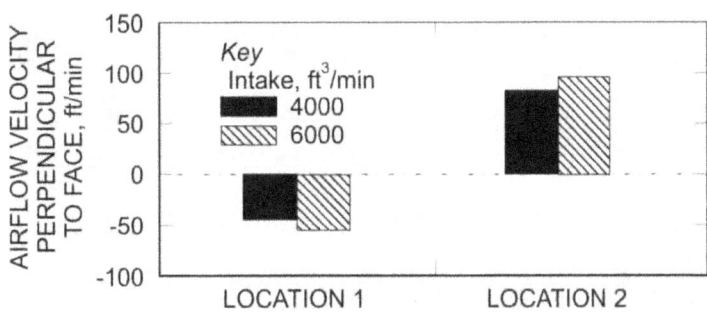

Figure 47. Airflow with no water sprays operating at locations 1 and 2 shown in Figure 45.

With the sprays operating, flows were measured at the curtain side (location 1) and off-curtain side (location 2) of the face (Figure 48).

Location 1: Curtain (intake air) side of the face

- The use of water sprays reversed the direction of the airflow on the left (curtain) side of the face (i.e., flow moved toward the face).
- Increasing the water pressure increased velocity.
- Velocities were higher when the angled sprays were used.
- Increasing intake airflow had only a small effect on face air velocity.

Location 2: Off-curtain (return air) side of the face

- Airflow direction, toward and away from the face, varied
  - With straight sprays, airflow moved toward the face.
  - With angled sprays, airflow moved away from the face.
- Increasing the water pressure increased velocity.

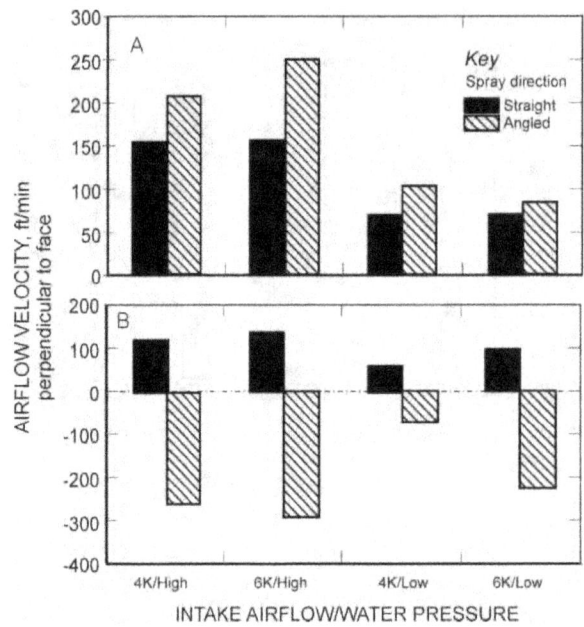

Figure 48. Airflow at curtain (A) and off-curtain (B) side of face.

The straight and angled sprays created different flow patterns at the face.

- With angled sprays
  o Air moved toward the face on the left (curtain) side of the entry and away from the face on the right (off-curtain) side.
  o Air moved left to right across the entire face.
- With straight sprays
  o Air moved toward the face on the right <u>and</u> left sides of the entry.
  o Air did not move across the entire face but moved over the top of the machine and away from the face. The movement of water mist over the machine gave evidence of this flow pattern.

Effect of Sprays on Methane Distributions

A series of tests were conducted to determine how water sprays affect methane concentrations above the mining machine in the area between the face and blowing curtain. Methane levels were measured at 15 locations above the mining machine (Figure 49), which was located at the center of the 13-ft wide entry. The sampling locations were 4 to 4½ ft from one another and 2 ft from the sides and face of the entry. Curtain setback was 35 ft. Intake flows were 4,000 and 6,000 ft^3/min.

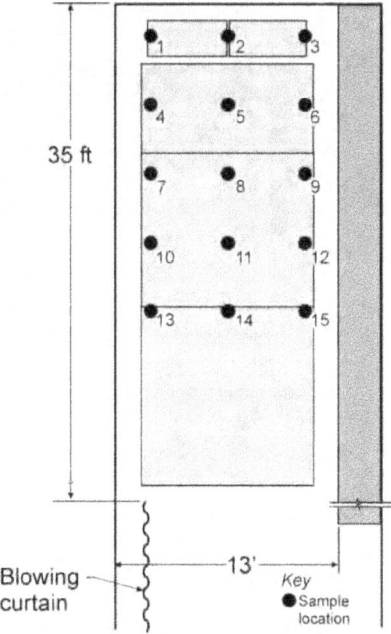

Figure 49. Methane sampling locations above mining machine (water spray tests).

The distributions of the methane over the mining machine, with and without water sprays, are shown in Figures 50 and 51.

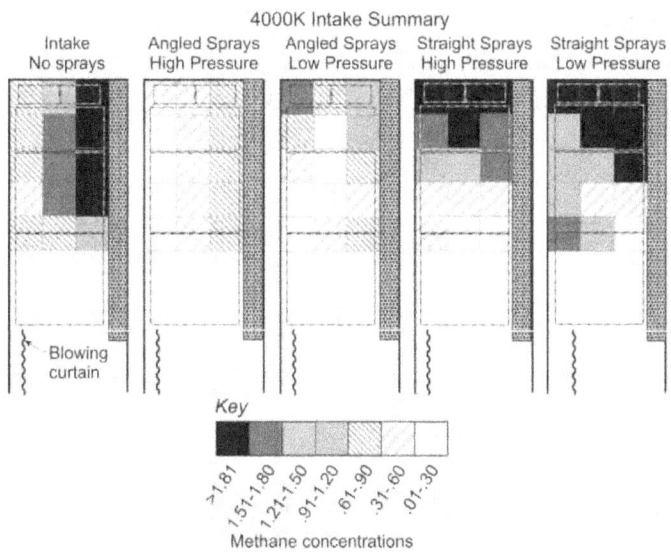

Figure 50. Methane distribution, 4,000 ft3/min intake flow.

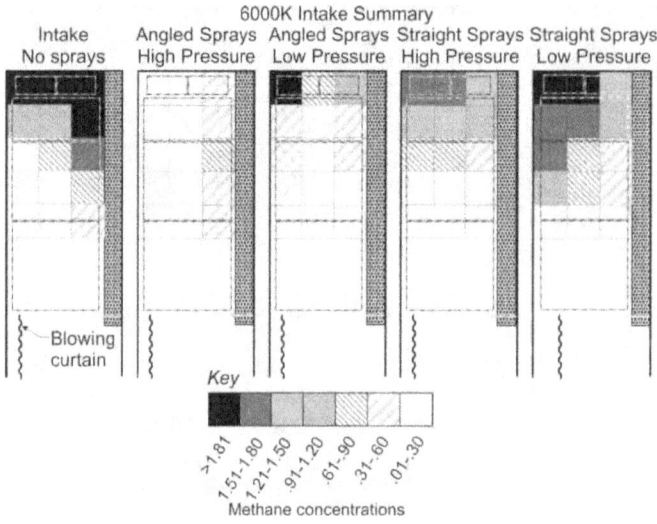

Figure 51. Methane distribution, 6,000 ft3/min intake flow.

- Water from the sprays angled to the right (return airflow) side of the face reduced methane levels at most of the sampling locations.
- Use of straight sprays with the lower intake flow resulted in higher methane levels near the face.

As methane gas moves further from a mining face it is diluted by the intake air. At some distance from the face, intake air and methane will be mixed uniformly and no further dilution will occur. To illustrate this progressive dilution of gas, the average concentration for each horizontal row of three samples was averaged and plotted versus distance from the face (Figure 52). Separate graphs are drawn to show how the concentrations varied with intake flow, spray direction, and water pressure.

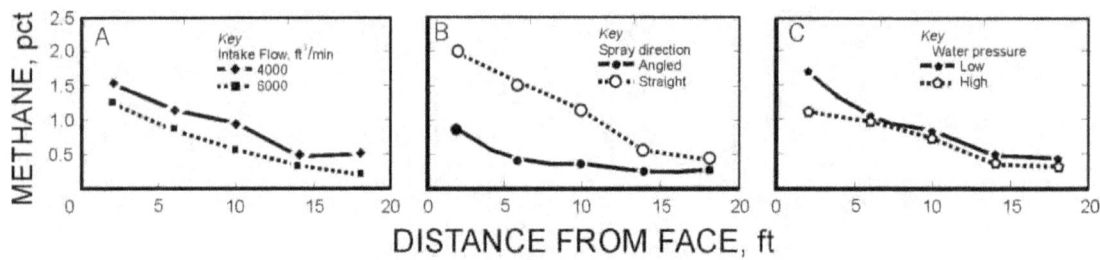

Figure 52. Effect of intake flow (A), nozzle direction (B), and water pressure (C) on concentration versus distance from face.

- Increasing intake airflow uniformly reduced methane levels as distance from the face increased (A).
- Angled sprays reduced methane levels more than straight sprays for distances up to 14 ft. Beyond 14 ft; the effect of spray type on methane dilution was negligible (Figure 52-B).
- Higher water pressure reduced methane levels but only at the face. At further distances, the water spray pressure had no effect on methane levels.

Summary

Operation of machine-mounted water sprays helps reduce face methane levels primarily within 5 ft of the face:

- Directing spray nozzles toward the return side of the face improves the removal of gas by increasing the velocity of airflow across the face.
- Directing water sprays straight toward the face will improve methane dilution but may result in higher methane levels until there is sufficient intake airflow to remove the gas from the face.
- The dilution of methane gas increases with distance from the face. Increasing water spray pressure improves dilution only within 5 ft of the face.
- Operation of water sprays does not increase the quantity of air reaching the face.

Note: These tests did not consider the effects of water spray operation and intake flows on airborne dust levels and, therefore, should not be used alone to develop a ventilation system for dust control.

METHANE MONITORING

Introduction

Engineering controls such as water sprays and scrubbers help maintain safe methane levels in coal mines. Methane monitoring is required to ensure that engineering controls are effective and that methane concentrations do not exceed regulatory standards. NIOSH research examined instruments and sampling strategies used for methane sampling and developed methods for evaluating performance. The research included monitors using both catalytic heat of combustion and infrared absorption sensors.

Methanometers, Catalytic Heat of Combustion Sensors

Figure 53 shows a catalytic heat of combustion sensor with its sensor elements. The catalytic sensor includes a very fine platinum wire contained within an alumina bead coated with a catalyst material, typically platinum or palladium. During operation, the bead (active) is heated to a temperature sufficient to promote combustion of the methane gas with oxygen on the surface of the catalyst. The increased heat generated by the combustion increases the resistance of the wire inside the bead. The change in resistance is monitored with a Wheatstone bridge circuit, which generates an electrical signal proportional to the methane concentrations. A second bead (inactive) in the bridge, not treated with the catalyst material, is used to compensate for changes in temperature, pressure, and humidity.

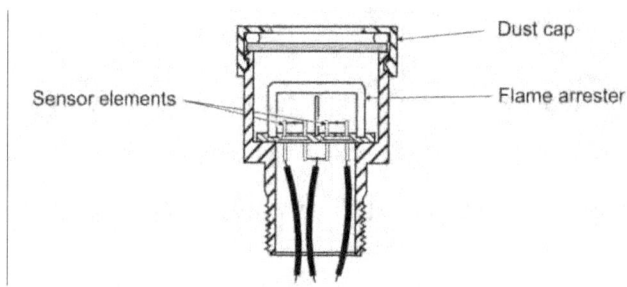

Figure 53. Sampling head with catalytic heat of combustion sensor.

The sensor head must be covered with a dust cap to protect the sensor element or sensor chamber from dust or water [30 CFR 27.22]. Within the cap, a flame arrestor made of screen or porous material is used to prevent any ignition of methane gas from moving outside the sensor head.

Methanometers, Infrared Sensors

Methanometers with infrared sensors operate on the principle that methane gas will absorb infrared light at certain wavelengths (e.g., 1.33, 1.66, 3.3, and 7.6 microns). For the instruments tested, a tungsten filament bulb provided an infrared light beam that is transmitted through an enclosure/cell that contains the sampled gas (Figure 54). A mirror at the end of this enclosure reflects the light back to a detector. An optical absorption filter placed in front of the detector limits the infrared light to a narrow band of wavelengths that are absorbed by the gas being measured (in this case methane). The detector measures the intensity of the filtered light, which varies inversely to the concentration of the methane in the enclosure.

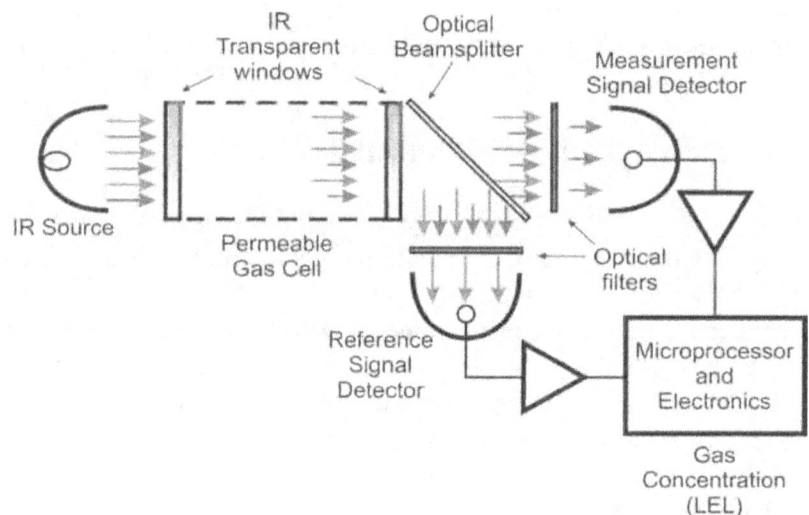

Figure 54. Components of infrared sensor.

Methanometers, Use in Underground Coal Mines

Methane monitors used in coal mines must be tested and approved by MSHA [30 CFR 75.342]. The design and use of approved monitors depends partly on whether they are designated as "methane monitors" or "hand-held detectors."

The methane monitor

- Is permanently mounted on a mining machine to provide continuous readings of methane levels near the face.
- Is located as close to the face as practicable, but at least 12 inches from roof, face, ribs, and floor [30 CFR 75.342(a)(3) and 30 CFR 75.323(a)].
- Provides a warning signal to alert workers whenever methane concentrations reach 1.0% [30 CFR 75.342(b)(1)].
- Provides an alarm signal to warn workers whenever methane concentrations reach 2.0% [30 CFR 75.342(c)(1)].
- Includes an electrical relay to cut power to the mining machine whenever methane concentrations reach 2.0% [30 CFR 75.342(c)(1)].

The methane detector is

- Battery powered and portable.
- Used to make 20-minute periodic gas checks at each active mining face [30 CFR 75.362)(d)(1)(iii)] and at outby locations.
- Used to make readings at least 12 inches from the roof, face, ribs, and floor [30 CFR 75.323(a)].

All methane monitors and detectors must be calibrated at least once every 31 days [30 CFR 75.342(a)(4)] according to procedures specified by the monitor manufacturer. All procedures use a standard gas (usually 2.5% methane by volume). Periodic bump tests should be performed between instrument calibrations, which involve exposing the sensor to a calibration gas and recording the concentration displayed on the instrument's visual readout. Typically, the displayed value should not vary more than 0.2% from the actual concentration of the calibration gas.

Measuring Methanometer Response Times

Instrument response time is defined as the time interval between the application of a gas of uniform concentration (e.g., calibration gas) to the sensor head and the final steady state reading of the instrument. Since the sensor output approaches the final reading asymptotically, the 90% response time[1] is often used to compare instrument performance. Currently there are no regulations regarding measurement of response times for instruments used in coal mines.

While a mining machine is cutting coal, methane concentrations at the face can rise and fall rapidly. The methanometers should not only monitor methane levels accurately but also be able to respond quickly to changes in methane concentration. If the methanometer response time is slow, the actual concentrations may be higher than the indicated readings.

NIOSH developed and tested two techniques for measuring methanometer response times for machine-mounted methane monitors [Taylor et al. 2002a, 2002b]. One response measurement

---

[1] *Time required for the instrument to read 90% of the final gas concentration*

technique was designed for use underground. It requires only a stop watch and equipment normally used for calibration. Methane gas is applied to the sensor head through the calibration cup. The methanometer does not have to be removed from the mining machine. The second is a simple laboratory technique that requires the methane sensor be placed in a test box where it is exposed to methane gas.

Underground Response Time Measurements Using a Calibration Cup

- Methane calibration gas (typically 2.5% by volume) is delivered at a constant flow rate through the manufacturer-supplied calibration cup.
- At time zero, the calibration cup is placed over the sensor dust cap.
- The elapsed time and corresponding concentration displayed on the visual readout are recorded every 5 seconds until the instrument readings become steady.
- The response time curve is drawn and used to determine the 90% response time.

Laboratory Response Time Measurement Using a Test Box

The wooden test box, 14 x 14 x 14 inch, is shown in Figure 55.

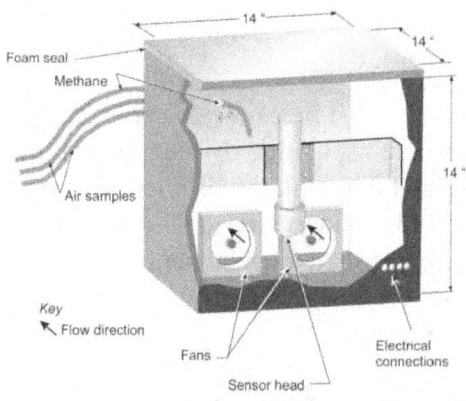

Figure 55. Test box used for response time measurements.

- One or more instrument sensors are placed in the box. The box is sealed by replacing the lid. The visual displays for each instrument remain attached to the sensors but are placed outside the box where they can be observed.
- One thousand cm^3 of methane gas (99% by volume) is injected into the box using a 1500-cm^3 graduated plastic syringe. Approximately 4 seconds are required to inject this quantity of gas. Two small fans mounted on the bottom of the box provide rapid mixing of the methane and air. Although there is some leakage from the box during and following methane transfer, the gas concentration remains approximately 2.2% by volume for more than 2 minutes.
- Time zero is when the gas is injected. Every 5 seconds, the gas concentration is read and recorded from the visual display until the readings no longer increase.
- Response time curves are drawn and used to determine the 90% response time.

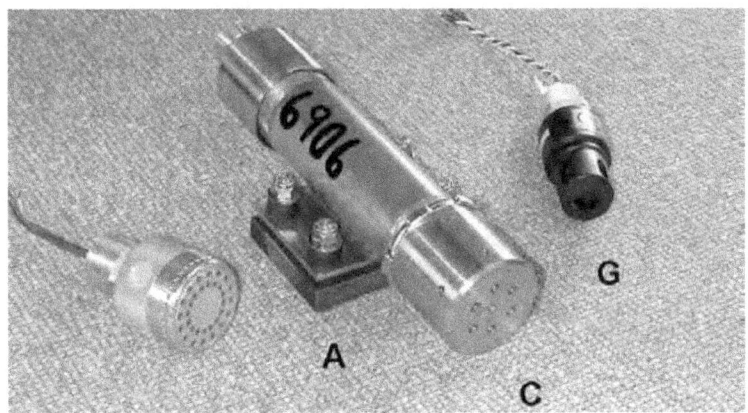

Figure 56. Sensor heads for monitors A, C, and G.

Measurement of Response Times, Machine-mounted Monitors

Three methane monitors that are approved by MSHA for use on mining machines were selected for testing. The three, identified as monitors "A," "C," and "G," (Figure 56) use catalytic heat of combustion sensors. The 90% response times were determined using the response time curves obtained using the two measurement techniques—calibration cup and test box—described above (Figure 57).

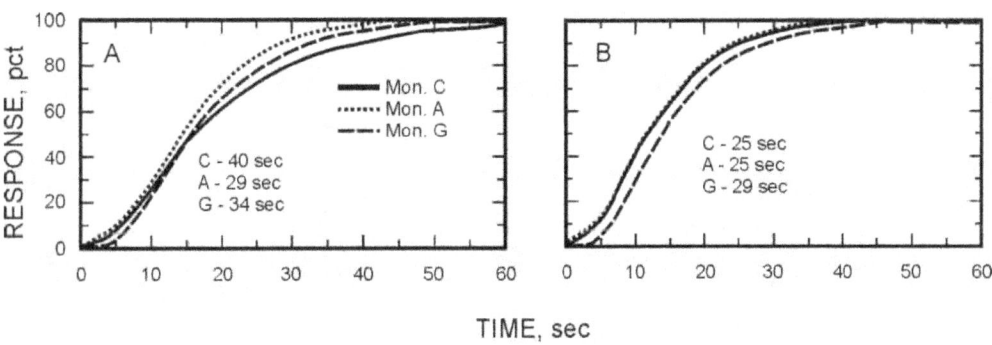

Figure 57. Response times measured with calibration cup (A) and test box (B).

- Response times varied between monitors with both measurement techniques. Flow patterns, which were affected most by the dust cap design (calibration cup test), had the biggest effect on response times (Figure 57 and 58).
- Placing the calibration cup on the dust cap restricted flow through the sensor head, especially with monitor C.
- Response times measured with the test box were shorter than when measured with the calibration cup.
  - In the test box, flow was not restricted by the calibration cup.
  - In the test box, flow through the sensor head was more representative of flow conditions underground.
- Using the test box, response times were measured with the dust caps removed from the sensor heads (Figure 59).

The response times were 6 to 16 seconds shorter when the caps were removed from the sensor heads. This represents the time required for the gas to diffuse through the dust cap.

Measurement of Response Times, Infrared Monitors

Most methane monitors used in coal mines have catalytic heat of combustion sensors, but some infrared monitors and detectors have been approved by MSHA for underground use. Response times were measured for two methanometers equipped with infrared sensors [Taylor et al. 2008]. The two sensor heads with their dust caps, designated IR-1 and IR-2, are shown in Figure 60. Both dust caps have narrow slot openings with baffles, which reduce the amount of dust and water mist that passes into the sensor. In addition, the inner surface of the IR-2 sensor head is lined with a filter material that resembles dense foam (180-micron pore size).

Response times were measured in the test box with and without the dust caps attached to the sensor heads (Figure 61).

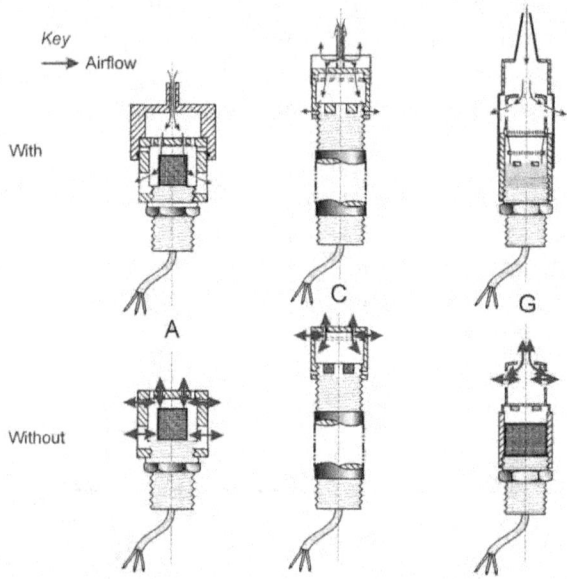

Figure 58. Sensor heads (A, C, and G) with and without the calibration cups attached.

- With the dust caps, the response time was three times longer for IR-2. The differences in the response times are due to the structure of the dust caps.
- Without the dust caps, the response times for both IR-1 and IR-2 were approximately 3 seconds. (Near the beginning of the test, concentrations exceeded 100% of the final response concentration—approximately 2% by volume—because methane gas injected into the box had not mixed completely).

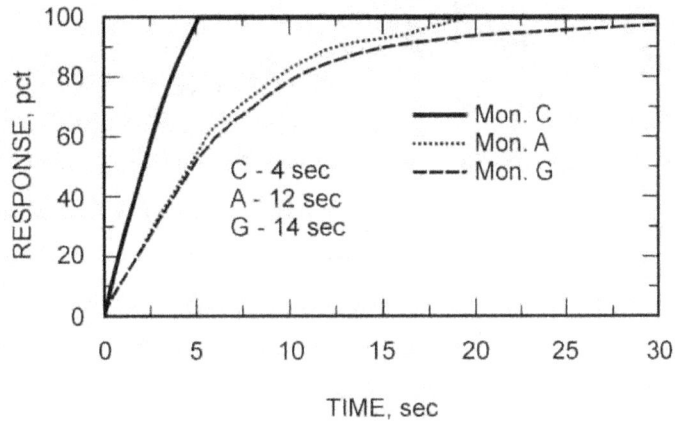

Figure 59. Response times measured in test box (dust cap off).

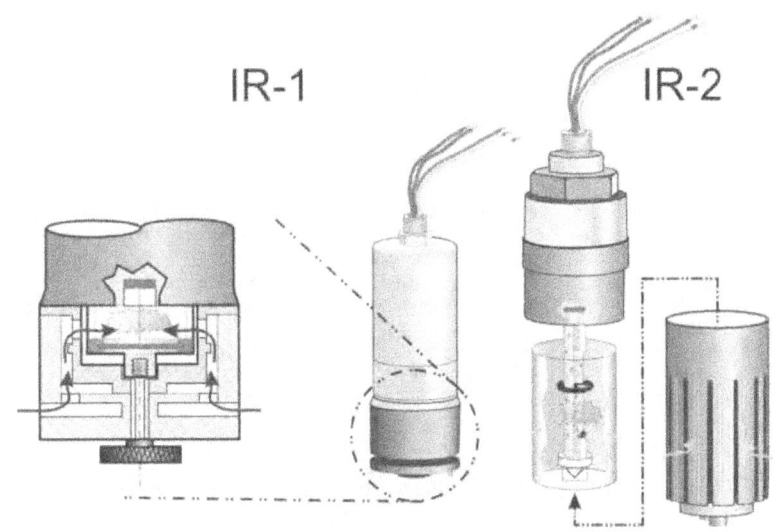

Figure 60. Infrared sensor heads with dust caps.

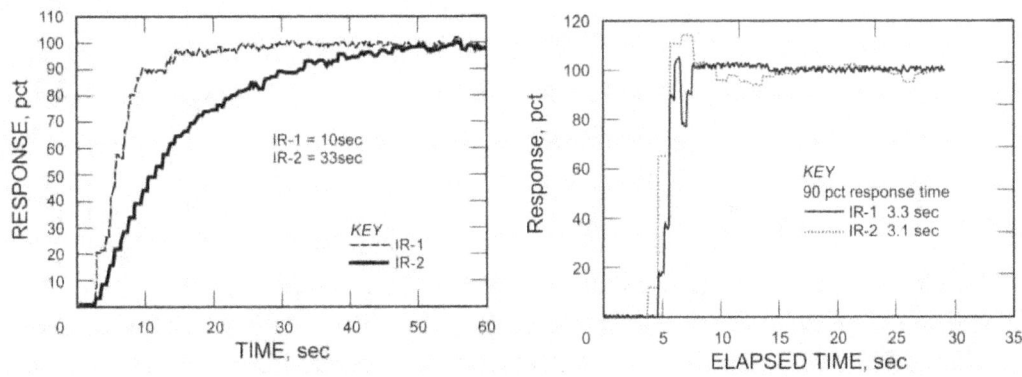

Figure 61. Response times with dust caps on (left plot) and off (right plot).

Effect of Response Time on Peak Concentration Measurements

A study was conducted to evaluate how response time affects face methane measurements.

47

Methane concentrations were measured with three instruments placed on the mining machine in the ventilation test gallery (Figure 62). Two of the methanometers (IR-1 and IR-2 described above) used infrared sensors and the third (HC) used a catalytic heat of combustion sensor. The heat of combustion instrument (Figure 63) was approved for use underground but not included in earlier response time tests. The 90% response times for the three instruments were

- IR-1–10 seconds
- IR-2–33 seconds
- HC–19 seconds

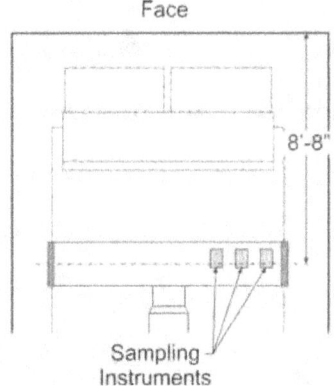

Figure 62. Sampling locations on model mining machine.

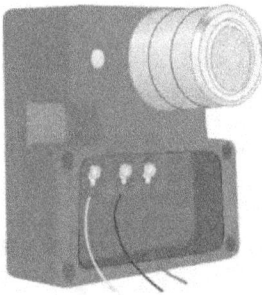

Figure 63. Catalytic heat of combustion sensor head.

Data were collected at a rate of one sample/sec for 2 minutes for several different ventilation conditions. Data from one of these tests are shown in Figure 64.

- The average concentrations for the entire sampling period measured by IR-1, IR-2, and HC were approximately the same (0.37%, 0.38%, and 0.41% methane respectively).
- Concentrations varied similarly during the sampling periods.
- The faster the instrument response, the quicker the concentration changed (up or down).
- The faster the instrument response the larger the peaks and dips in concentration.

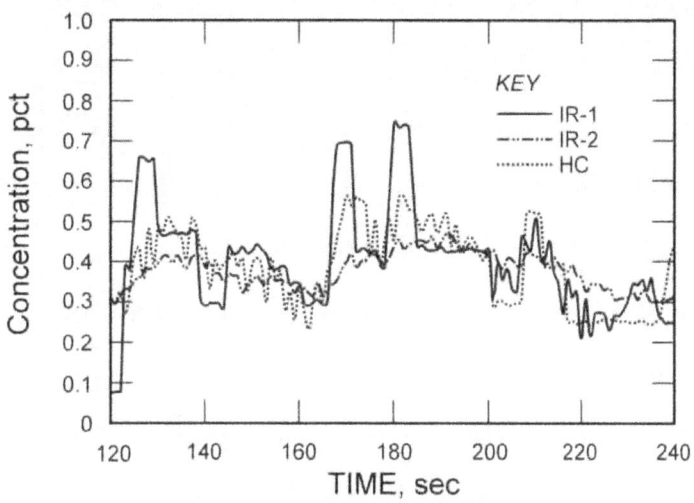

Figure 64. Data collected using three methanometers.

Methane Sampling Strategy, Sampling on the Mining Machine

Methane measurements are made on the mining machine to estimate face methane concentrations. Frictional ignitions are most likely to occur at the face where it is not possible to measure methane concentrations during mining. As long as methane concentrations measured on the machine are less than 1%, methane concentrations at the face are assumed to be less than 5%, the lower explosive limit for methane. Whenever concentrations measured on the machine exceed 1%, the protection provided to the worker is reduced.

Sampling location on the mining machine is one of the most important factors affecting the estimates of face methane concentration and protection provided to the worker. Federal regulations require that "...the [methane] sensing devices of methane monitors shall be installed as close to the working face as practicable [30 CFR 75.342(a)(3)]." The mining company, with the approval of MSHA, usually selects the location for mounting the methanometer sensor.

Common guidelines for sensor placement include locating the sensor

- Six to 8 ft from the face where damage to the head due to falling rock and moisture is less.
- On the return air side of the mining machine (side opposite the ventilation tubing or curtain) where methane concentrations are usually highest.

Tests were conducted in the ventilation test gallery to compare methane concentrations measured at several locations on the machine with methane concentrations measured concurrently at locations 1 ft from the face. For these tests, the mining machine was located at the center of a 16½-ft wide entry. Blowing tubing, positioned 2 to 10 ft behind the mining machine, directed intake air (4,000 or 7,000 ft^3/min) down the left (intake air) side of the entry. Machine-mounted sprays and a scrubber were operated during these tests.

Methane sampling locations were 5.5, 6.5, and 26 ft from the face, on both sides of the model mining machine. In addition, there were three sampling locations 1 ft from the face (Figure 65).

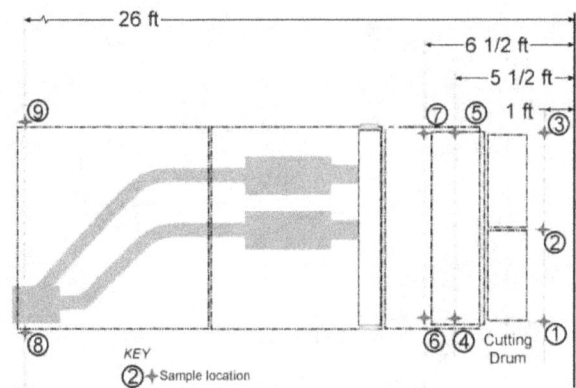

Figure 65. Methane sampling locations on mining machine and at face.

Concentrations were monitored at all locations on the machine and at the face for eight different face ventilation conditions and tests were repeated once. For each test, the highest of the three face concentrations was plotted versus the average concentration measured at each of the six machine sampling locations. Scatter plots with the linear lines of best fit for each sampling location are given in Figure 66.

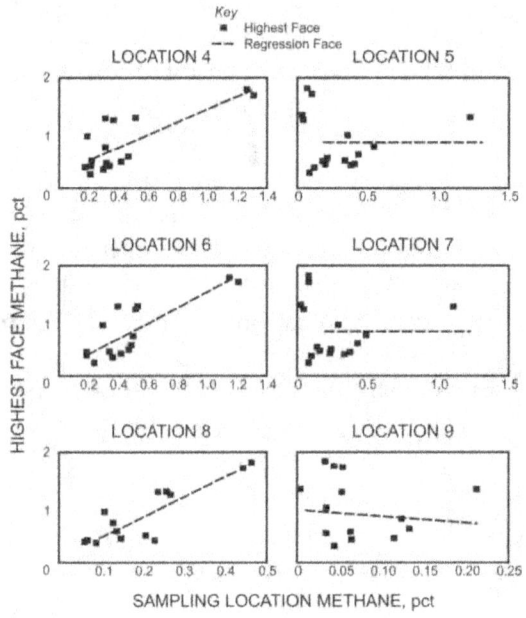

Figure 66. Comparison of measurements on mining machine. Locations given in Figure 65.

The "student's t" distribution was used to determine if the slope of each line was statistically different from zero (95% confidence). Statistically, the relationships between machine and face concentrations were significant only when the slope of the least squares line was greater than zero. Analysis of the data showed

- The relationships were significant only for the three sampling locations on the right (return air) side of the machine (locations 4, 6 and 8): airflow moved away from the face, passing over the three locations (4, 6, and 8) on the right (return air) side of the face.

In theory, all three locations on the right (return air) side of the machine could be used for sampling, but

- Location 8 would probably not be chosen because it is 26 ft from the face.
- Location 4 would probably not be chosen because it is in a location (5.5 ft from the face) where damage to the sensor is likely.
- Location 6 is relatively close to the face (6.5 ft) but far enough to provide protection for the sampling instruments.

The relative protection provided to the worker was evaluated by sampling at two locations on the return air side, 5 and 6.5 ft from the face (locations 4 and 6, respectively). First, it was assumed that, since location 4 was closest to the face, measurements made at this location provide the best estimate of face concentration. The 95% prediction limits were added to the line of best fit for location 4 (Figure 67). Based on this data, if the measured concentration is 1%, the chance of the highest face concentration exceeding 2.3% is less than 5%. If this is an acceptable level of risk, an equivalent level of risk can be determined for measurements made at location 6.

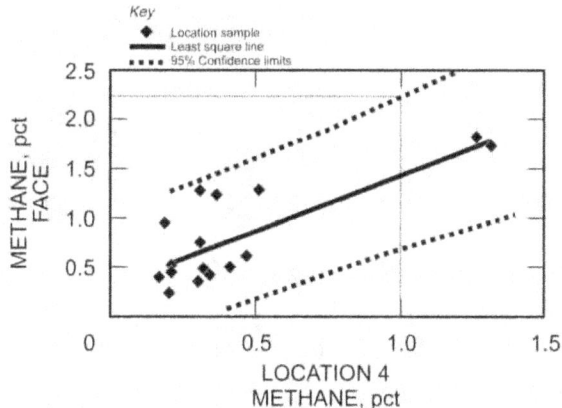

Figure 67. Comparison of face methane concentrations versus concentrations at location 4 (return air side, 5.5 ft from the face).

Assuming there is a linear relationship between concentrations measured at locations 4 and 6, a concentration of 1.0% measured at location 4 corresponds to a concentration of 0.96% measured at location 6 (Figure 68). Equivalent levels of protection would be provided if the maximum allowable concentrations at locations 4 and 6 were 1.0% and 0.96%, respectively.

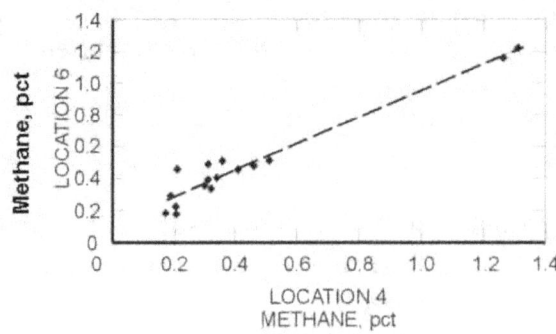

Figure 68. Concentrations measured at locations 4 and 6 (return air side, 5 and 6.5 ft from face, respectively).

Summary

Methane monitoring is needed to verify that methane levels do not exceed regulatory standards. This section examined methanometers using catalytic heat of combustion and infrared technologies and evaluated methods for assessing the response times of the various sensors.

- The techniques developed by NIOSH for measuring response times provide a means for comparing the performance of methanometers used in coal mines.
- Response times for catalytic heat of combustion and infrared sensors varied primarily due to the designs of the dust caps.
- Dust cap design and not sensor type (infrared or catalytic heat of combustion) is more important in determining response time.
- Although instruments with faster response times provide more accurate measurements of methane concentration, the sensor heads must be protected from exposure to excessive amounts of dust and water spray. The amount of protection needed for each type of sensor to perform accurately can only be determined by long-term, underground tests.
- Methane sampling location on the mining machine is critical for accurately estimating face concentration and establishing an acceptable level of safety.

MEASURING GAS LEVELS OUTBY THE FACE

Introduction

Regular monitoring for methane gas is required near faces where methane concentrations are usually highest and the potential for methane ignitions is the greatest. However, methane gas can also accumulate in areas outby the face where methane concentrations are not monitored on a regular basis.

It is not practical to place sampling instruments in all areas where miners work and travel. Providing methane monitors that can be worn by the miner is one way to protect workers regardless of their work location.

Person-Wearable Methane Monitors

Although personal monitoring for methane is currently not required in U.S. coal mines, monitors used to protect workers from exposure to high concentrations of combustible gases, such as methane, are used by in many non-mining industries. Personal combustible gas monitors were obtained from seven different manufacturers [Chilton et al. 2005]. The monitors (designated A through G)

- Were small and designed to be worn on the worker's clothing.
- Equipped with a digital readout that displayed methane concentration.
- Had visual, audible, and vibratory alarms to alert workers when the methane levels exceed preset limits.
- Had catalytic heat of combustion sensors that could measure methane levels up to 2.5% by volume.
- Were all approved by MSHA for use in hazardous locations other than coal mines, but at the time of testing none of the instruments were approved for use in coal mines.

The performances of seven of these instruments were compared by measuring instrument drift with time and instrument response time. One additional monitor, incorporated into a miner's cap lamp, was evaluated separately. The operational characteristics of the instrument alarms were also compared.

Instrument Drift

The instruments were calibrated before beginning the tests. The seven instruments were placed in the 14- x 14- x 14-inch test box described earlier. A clear plastic top used to seal the box allowed the visual displays to be observed. The instruments were operated 8 hours a day for 12 consecutive workdays.

Twice each day, morning and afternoon, 1,300 ml of 99% methane gas was injected into the box using a 1,500-ml syringe. Although some gas leaked from the box during and following the transfer of gas, the resulting concentration in the box remained relatively constant (approximately 2.5% by volume) for more than 2 minutes. Concentrations were read from the visual displays 1 minute after the gas was injected. The daily readings are shown in Figure 69.

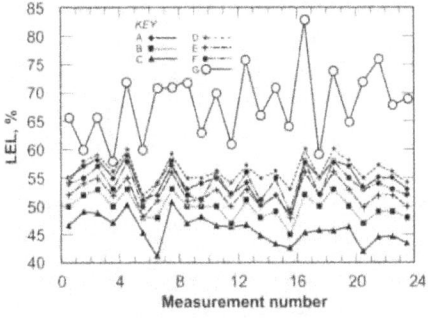

Figure 69. Daily readings for person-wearable monitors.

- Five of the seven instruments gave readings that were within variations allowed by MSHA for portable gas detectors.
- Readings from six of the seven monitors consistently tracked one another during the 12 days of testing. The differences in the readings were most likely due to calibration procedures used for individual instruments. Each instrument manufacturer supplies a calibration cap made for that instrument. The cap design affects how calibration gas flows over the sensor head and the calibration setting.
- One instrument gave readings considerably different from the other six instruments.
- To assure accuracy, calibration must be performed at least once every 31 days. Bump tests should be conducted more frequently.
- The amount of instrument drift would be expected to increase with instrument age and exposure to dust and water spray.

Instrument Response Time

Instrument response time was measured; one instrument at a time, in the test box using the procedure described earlier [Taylor et al. 2002a]. The response time curves are shown in Figure 70.

- The 90% response times varied from 8 to 20 seconds. These times were faster than response times measured for machine-mounted methane monitors. However, the person-wearable devices have limited filtering capability for removing airborne water and dust.

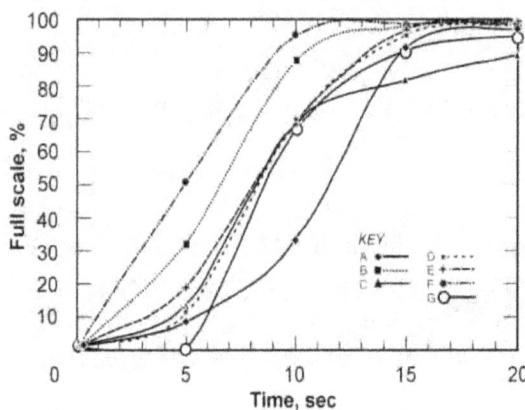

Figure 70. Response times for person-wearable monitors.

Instrument Alarms

Each of the seven methanometers tested has visual, audible, and vibratory alarms that produce signals to warn the worker when gas concentrations exceed preset limits. There were two alarm limits that could be set to signal for two separate methane levels.

Visual: Visual alarms were located at one or more locations on the instrument.

- All visual alarms had similar light intensities except instrument B which was more difficult to see.
- All used pulsed lights to improve recognition. Pulse frequency varied for high and low alarms.
- The effectiveness of the visual alarms will be highly dependent on the locations of the lights (e.g., top or side of the instrument) and where the instrument is located on the individual.

Audible: Audible alarms were produced by a speaker located in either the front or top of the methanometer.

- The maximum A-weighted sound levels ranged from 72 to 88 dB for the top microphone location. Since ambient A-weighted sound levels near operating machinery in underground mines are generally in excess of 85 dB, the audible alarms would probably not be heard by miners working in face areas. The audible alarm probably could be heard by persons working in areas outby the face.

Vibration: Vibration alarms were produced by a motor that rotates an eccentrically weighted shaft. The vibrator was located either in the instrument itself or in a detachable unit.

- Vibratory frequencies ranged from 80 Hz to 160 Hz, and displacements ranged from approximately 5.4 to 22.2 µm.
- The closer the vibrator is kept to the skin the more likely the vibratory alarm will be recognized. Instruments with detachable vibrators are probably more effective because the vibrator can be placed closer to the skin while the monitor can be attached to outer clothing.

Positioning of the personal monitors is very important.

- The instruments should be attached to workers clothing at locations where air can freely pass over the sensor head.
- Sampling instruments should be positioned at locations where the miner can readily identify one or more of the alarms (visual, audible, and vibration).

Cap Lamp-Mounted Personal Monitor

A personal monitor that uses a cap lamp-mounted methane sensor was also tested (Figure 71) [Chilton et al. 2003]. The monitor is powered by the cap lamp battery and continuously monitors methane as long as the cap lamp is operating. Audible and visual (buzzer and blinking cap lamp) signals are initiated when methane levels exceed preset limits.

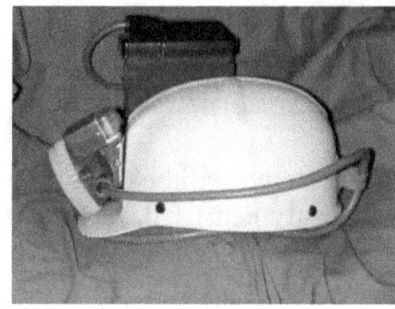

Figure 71. Cap lamp-mounted methane monitor.

Since this monitoring device has no visual readout, a test protocol had to be developed for measuring instrument accuracy. Evaluations were based on the time between application of the calibration gas through a calibration cup to the sensor head and the initiation of the alarm signals. Signal amplification was adjusted using a potentiometer so that the alarm time for 1% calibration gas was between 15 to 25 seconds. When the alarm times were within this range the instrument was considered calibrated.

After calibration with 1% calibration gas, the response of the instrument to other concentrations was determined. Calibration gases (0.6% to 2.5% by volume) were applied to the sensor heads using the calibration cup, and the time for obtaining an alarm signal were recorded (Figure 72).

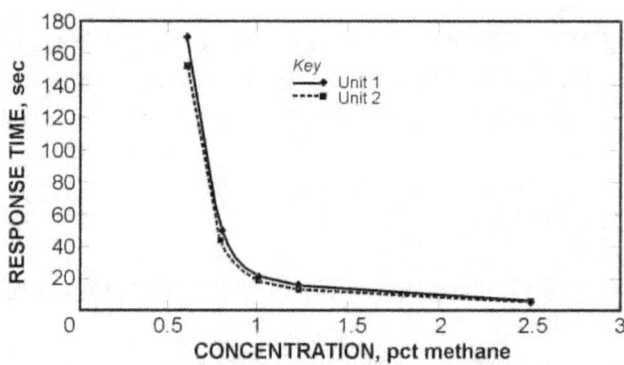

Figure 72. Instrument alarm times (calibration gas 0.6% to 2.5%).

- Alarm times increased with decreasing methane concentrations and decreased with increasing concentrations.
- Below a "threshold concentration" (e.g., less than 0.8% methane) alarm times were much longer (greater than 1 minute).

After calibration, two instruments were operated for 10 consecutive workdays. Twice each day, 1% calibration gas was applied to the sensor heads. The resulting response times are shown in Figure 73.

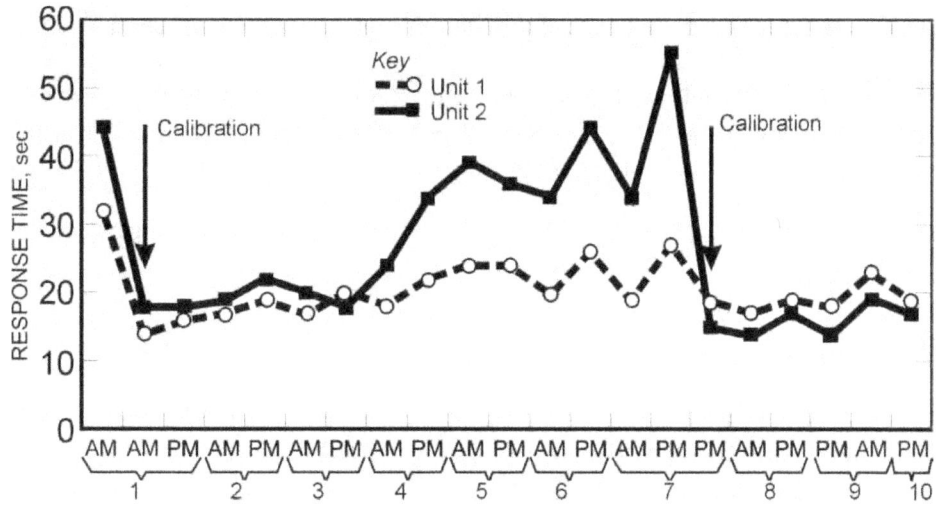

Figure 73. Instrument readings for 10 workdays.

- One instrument displayed little drift in response time, but time for the second to alarm increased significantly after 3 days. This second instrument was recalibrated after 7 days.
- The cap lamp-mounted methane monitor provides a convenient way to continuously monitor methane levels in all areas where a miner works. However, without a visual readout to show concentrations, checking the instrument performance is more difficult.
- The alarms were difficult to recognize.
    - The cap lamps blink rapidly when signaling an alarm. The rapid blink rate can be detected by a second individual. But the person wearing the lamp may not detect the signal, especially if the worker is moving and/or rapidly moving his or her head.
    - The aural signal was 60 to 68 dBA at a distance of 1 ft from the instrument. It is unlikely the alarm could be heard when working near mining equipment having background noise levels of 85 dBA or higher.

Summary

Research was conducted to examine if person-wearable methane monitors could be used to provide warnings when high methane concentrations occurred in areas outby the mining face. With the instruments, alarm signals can be set to two levels, and each has visual, audible, and vibration alarms. Criteria were established for comparing the effectiveness of the warning signals. For most of the person-wearable instruments, performance (accuracy and response times) was comparable to currently approved handheld methane detectors.

EVALUATING METHANE LEVELS IN AREAS OUTBY THE MINING FACE

Introduction

The effects of ventilation changes on methane levels outby the face are not usually known because methane sampling is not required in most outby areas. However, it is assumed that gas levels will increase in all areas of the mine when a main mine fan stops operating. Knowing how quickly, and under what circumstances, the levels may increase will help emergency planning.

As a safety precaution, miners must begin evacuating the mine within 15 minutes of a fan stoppage. During evacuation, it is important that gas levels remain below the lower explosive limit along the haulage ways where the miners must travel, at least until they reach the portal. If methane concentrations in the haulage ways are greater than 5% by volume, and electrically powered vehicles are used to evacuate the miners, a spark from one of the electric motors could ignite the methane gas.

Underground Tests

Tests were conducted in four underground mines during planned fan stoppages to determine where and how much methane levels change in haulage ways immediately after the stoppages [Taylor et al. 2002b]. During the stoppages, methane concentrations were monitored continuously at 5 to 10 locations along the haulage ways and in areas near one or two of the active faces. The haulage way locations were either in or near entries that would be used by miners when leaving the mine.

Selection of the sampling locations was based on recommendations by mine personnel where they expected methane levels to be highest during a fan stoppage. Some of these areas were near worked-out, but ventilated, areas or old workings that had been sealed. In each selected area, sampling instruments were placed close to high points in the roof or where air velocities would normally be lowest. Figure 74 shows the 11 selected locations of the sampling instruments in one of the mines.

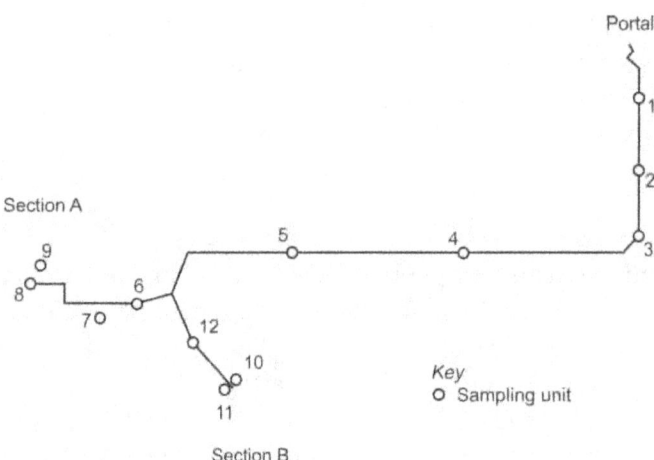

Figure 74. Methane sampling locations in one mine.

The instruments were placed at the sampling locations before the fan was turned off. MSHA-approved methanometers equipped with recording devices were used since no personnel were permitted underground during the fan stoppages. The duration of the fan stoppages was 3 to 5 hours. The instruments were retrieved from the mine after the fan was restarted and safety inspections were conducted.

- In three of the mines, methane concentrations along the haulage ways remained at or below the detection limit of the methanometer for the entire time the fan was shut down.
- At the fourth mine, all readings along the haulage ways remained zero except at two locations where concentrations reached as high as 0.7%.

In most of the sections, sampled methane concentrations reached higher levels (up to 1.2%) during the fan stoppages than in areas along the haulage ways. Figure 75 shows the concentrations measured near the face of one of the sampled sections. Concentrations at the face gradually increased to 0.4% following the fan stoppage, which lasted about 3 hours. After the fan was restarted methane levels peaked in the immediate returns. The methane concentration in the return was still increasing 90 minutes after the fan was restarted.

Although the concentrations measured at most of the sampling locations were very low, methane gas was continuously accumulating at the sampling locations. However, assuming the miners would begin evacuation of the mine within 15 minutes of the fan stoppage, it is unlikely they would have been exposed to concentrations greater than 0.2%. At the four mines where measurements were made, use of electrically powered haulage to evacuate the mine would not have significantly increased the risk of a methane ignition.

These results were based on data collected in four mines. The results show that methane concentrations will increase in underground workings during fan stoppages, but the size of the increase could vary significantly based on many factors including specific methane content of the coal and coal permeability. Decisions concerning evacuation procedures should be based on data collected at individual mines.

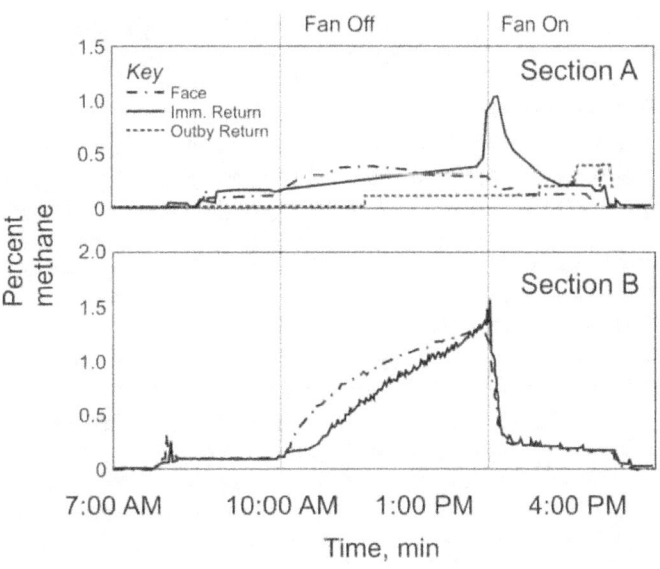

Figure 75. Accumulation of gas at two faces in one mine.

Summary

Methane concentrations were measured along haulage ways in four underground mines to measure methane concentrations during fan stoppages and to determine the safety of evacuating workers using electrically powered vehicles. Methane sampling instruments with data loggers were used to monitor methane levels during the fan stoppages, which lasted 3 to 5 hours. Although methane levels increased at some of the sampling locations, nowhere along the haulage ways did methane levels exceed 2% by volume. In the mines where tests were conducted, workers could safely use electrically powered haulage to evacuate the mine if they

- Began evacuation within 15 minutes after fan stoppage.
- Went immediately to the mine portal.
- De-energized circuits after the miners were withdrawn from the mine.

Individual mines should use sampling procedures developed by NIOSH to evaluate the safety of using electrically powered haulage to evacuate workers during a fan stoppage.

TECHNIQUES FOR MEASURING AIRFLOW

Introduction

Fresh intake air is needed at the face to dilute and remove methane emitted from the mining face. Effective ventilation requires that sufficient air be delivered to the face to dilute and remove emitted methane. Federal regulation requires that a minimum of 3,000 ft^3/min must reach each working face and this quantity must be measured "... at or near the face end of the line curtain, ventilation tubing, or other ventilation control device" [30 CFR 75.325(a)(2)]. For faces with exhaust ventilation, the mean entry air velocity must be measured "...at or near the inby end of the line curtain, ventilation tubing, or other ventilation control device" [30 CFR 75.326].

A study described earlier [Thimons et al. 1999] showed that airflow quantities measured at the mouth of the blowing curtain are usually not good estimates of how much air actually reaches the face. Better estimates of face airflow could be made closer to the face if airflow monitoring equipment was available. Vane anemometers must be aligned with airflow to give accurate readings and do not work well inby the tubing or curtain where air direction is constantly changing. The use of smoke for measuring flow at the face is usually not possible due to flow turbulence.

Ultrasonic Anemometers

Ultrasonic anemometers are used for collecting surface meteorological data. Accurate readings can be taken in areas where flow is turbulent and velocities are either high or low. One-, two-, and three-axis anemometers were obtained for use in the ventilation test gallery. Sensor heads for the three anemometers are shown in Figure 76.

The one-axis anemometer measures flow velocity in the direction of the instrument's orientation. The two-axis instrument measures flow velocity in the UV plane with a magnitude in the direction of flow equal to the square root of the sum of the squares of the velocity components. The three-axis instrument is also used to measure flow in the UV plane. In addition, the magnitude of the w vector, which is perpendicular to the UV plane, is recorded.

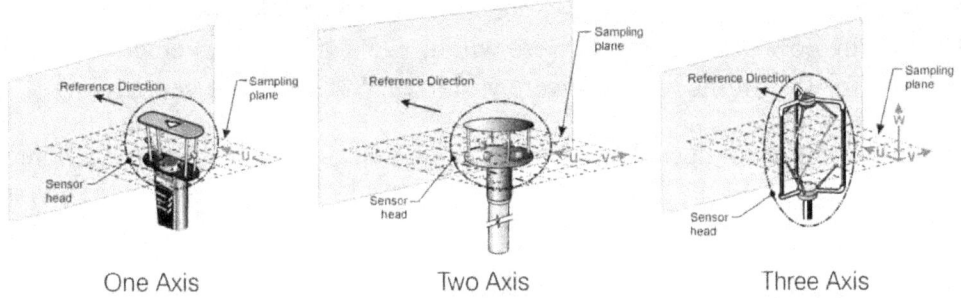

Figure 76. Ultrasonic sensor heads with flow components.

Evaluating Performance of Ultrasonic Anemometers

Airflow monitoring instruments are normally evaluated in a wind tunnel where airflow speed and direction can be precisely controlled. A suitable wind tunnel was not available initially and techniques were developed for evaluating anemometer performance at the NIOSH facility.

Low Air Velocity Measurements

The anemometer manufacturer states that instrument resolution is approximately 2 ft/min. NIOSH designed and built a low air velocity apparatus (LAVA) (Figure 77) specifically for testing instrument accuracy at air velocities as low as 2 ft/min [Hall and Timko 2005].

Most wind tunnels are designed to direct air over the test instruments at known and constant speeds and cannot be used for velocities less than 30 ft/min. With the LAVA, the instrument moves at a known constant speed through a tunnel in which air is not moving.

Figure 77. Low air velocity apparatus.

The LAVA tunnel is 26 ft long, 4 ft wide, and 4 ft high and sealed with plastic to minimize flow in and out of the tunnel. A cart moves along a straight track that runs the length of the tunnel. A synchronous motor operated with a programmable controller moves the cart, designed to hold a one-axis ultrasonic anemometer, along the track at a constant speed. The cart speed, therefore, is equal to the speed of the air passing over the anemometer.

Cart speeds were varied from 2 to 20 ft/min. Test results showed

- At 20 ft/min all anemometer readings were within ±25% of the true velocity.

- At 10 ft/min, 47% of the readings were within ±25% of the true velocity.
- At 5 ft/min, 46% of the readings were within ±25% of the true velocity.
- At 2 ft/min, 35% of the readings were within ±25% of the true velocity.

Tests in the LAVA were conducted only with the one-axis anemometer. All ultrasonic anemometers use the same type sensors and should provide similar accuracy for low air velocity measurements.

Comparing Instrument Airflow Readings in the Ventilation Test Gallery

Two identical three-axis anemometers (labeled A and B) were placed behind the curtain in the ventilation test gallery (Figure 78) and exposed to the same airflows [Taylor et al. 2003, 2002a]. The objective was to determine the variation in readings between the two instruments.

Velocities behind the curtain ranged from 100 to 700 ft/min. Flow velocities were adjusted by opening or shutting the regulator doors. A round bubble gauge was used to position the anemometers vertically. The sensor heads were placed at the center of the area behind the curtain and rotated so the reference arrows on the heads were aligned with the airflow direction. For each test, airflow data were collected for 3 minutes, the instrument positions were reversed, and the tests were repeated. Figure 79 shows the average velocities measured with the two instruments.

- Average readings varied by less than 5 ft/min for velocities of 410 and 700 ft/min.
- Average readings varied by less than 15 ft/min when the velocity was 140 ft/min.

An examination of the real time velocity data showed flow varied more for the lowest velocity. The effect of wind speed and direction outside the test gallery on ventilation inside the test gallery appeared to be greater when airflow entering the test gallery was less.

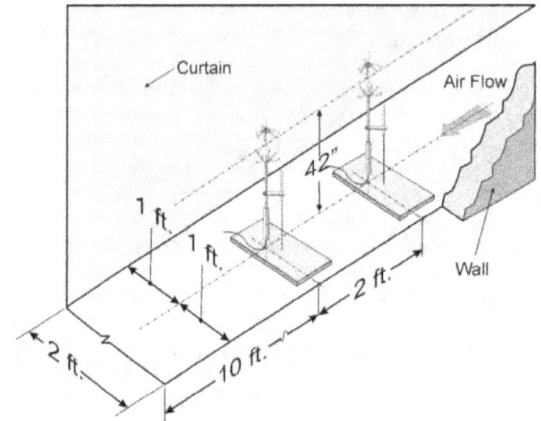

Figure 78. Airflow sampling locations behind the curtain.

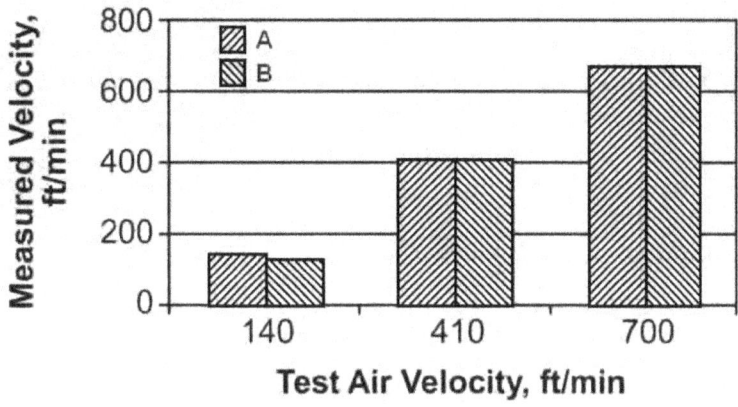

Figure 79. Velocity readings for three flows measured with two, 3-axis anemometers.

Comparison of One-, Two-, and Three-Axis Anemometers Behind the Blowing Curtain

In a second series of tests, one-, two-, and three-axis anemometers were individually placed at the same locations behind the curtain and exposed to the same airflows for 3 minutes. In each case, the instruments were oriented vertically with the sensor heads oriented in the direction of the airflow. The averaged velocities measured with each instrument are shown in Figure 80.

- One-, two-, and three-axis anemometers gave approximately the same readings when exposed to the same airflow behind the blowing curtain.

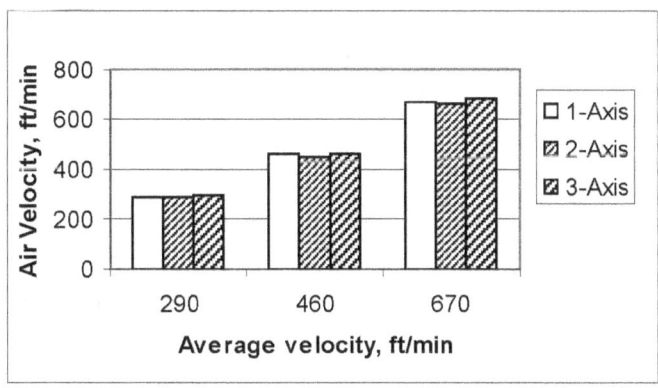

Figure 80. Velocity readings for three different flows measured with 1-, 2-, and 3- axis anemometers.

Instrument Orientation

For all measurements made behind the blowing curtain, the instruments were oriented vertically with reference arrows pointed toward the face. Tests were conducted to determine how instrument orientation affects flow velocity readings [Hall and Timko 2005].

Effect of Yaw Angle

Yaw angle is defined as the number of degrees an instrument is rotated in the clockwise direction from a reference direction (Figure 81). One-, two-, and three-axis anemometers were individually placed behind the curtain and exposed to an air velocity of approximately 400 ft/min. The first velocity measurement was made with the reference mark on the instrument pointed toward the face. Additional readings were taken as the instrument was rotated clockwise in 10-degree increments (Figure 82).

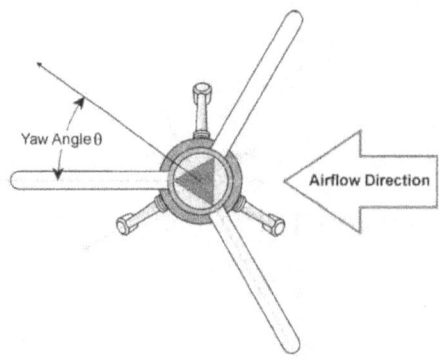

Figure 81. Measuring a yaw angle.

- Two- and three-axis velocity readings were unaffected by changes in yaw angle.
- One-axis velocity readings decreased when the yaw angle increased above 20%. The results were similar to those obtained by Boshkov and Wayne [1955] who, using a 3-inch rotating vane anemometer (a one-axis instrument), showed that yaw angles up to 20 degrees do not affect the registered velocity.
- Single-axis instruments should only be used to make flow velocity readings when flow direction is known and the instrument can be oriented in the direction of the flow.

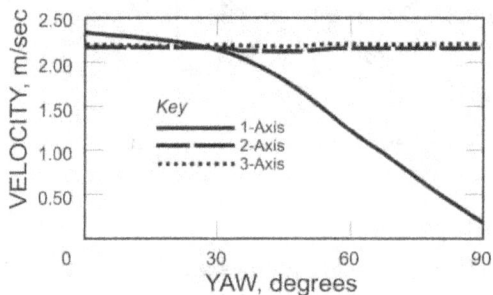

Figure 82. Effect of yaw angle on anemometer readings.

Effect of Tilt Angle

Tilt angle is defined as the number of degrees an instrument is slanted into the direction of the airflow (Figure 83). Starting from a vertical position, each anemometer was incrementally tilted to angles up to 90 degrees. The results are shown in Figure 84.

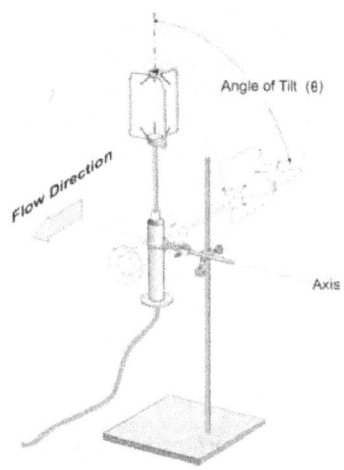

Figure 83. Evaluating effect of tilt angle on airflow readings.

- All velocity measurements decreased with increasing tilt angle, but the patterns were different for each instrument. The differences are believed to be due to the structures of sensor heads, which affect flow patterns. For each tilt angle, the flow is partially obstructed by the physical structure of the sensor head.
- All the instruments should be oriented vertically to minimize effects of tilt angle on readings. However, tilt angles less than 30 degrees will not have a significant effect on instrument readings.

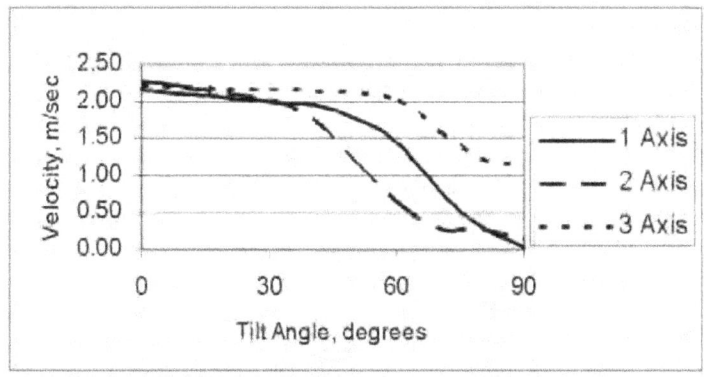

Figure 84. Effect of tilt angle on airflow readings.

Flow Readings with One-, Two-, and Three- Axis Anemometers at the Mining Face

Airflow direction behind the blowing curtain is relatively constant. Airflow between the curtain mouth and the face is constantly changing. A series of measurements were taken with the anemometers placed at three locations 2 ft from the face of the ventilation test gallery [Hall et al. 2007] (Figure 85).

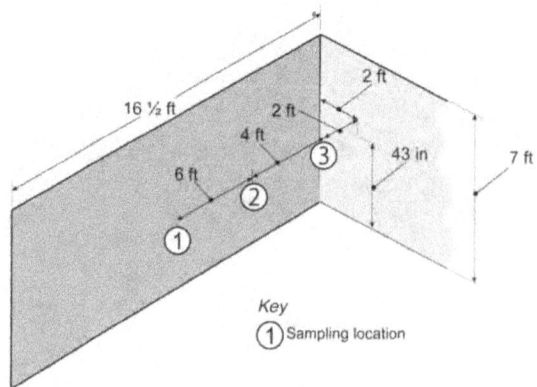

Figure 85. Face airflow sampling locations.

The instruments were positioned vertically and with reference arrows pointed toward the face. Air velocity measured at the mouth of the blowing curtain (35-ft setback) was maintained at 400 ft/min. The anemometers were rotated between the three sampling locations, and the tests were repeated. The average velocity readings for the three locations are shown in Figure 86.

- Measurements with the one-axis instrument were much lower than readings from the two- and three-axis instruments; and at location 3 on the right side of the cutting head at the face (Figure 85), measurements with the one-axis instrument were negative (indicating flow away from the face).
    - Velocity differences were due to alignment of the instrument with the face. The average direction of flow varied across the face (Figure 87). On the right (return air) side of the entry where flow was directed away from the face, the velocity measured by the one-axis instrument was negative.

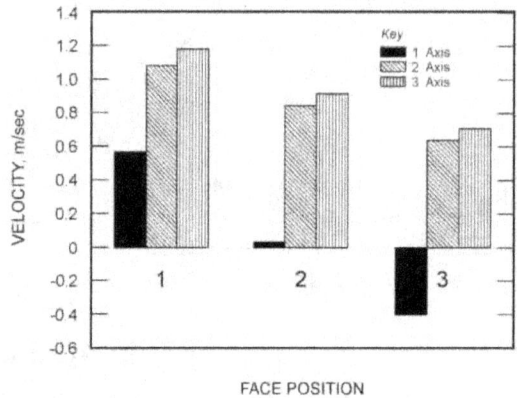

Figure 86. Anemometer readings at face sampling locations.

- Velocities measured with the two- and three-axis instruments were similar, but differed more than the readings taken behind the curtain.
    - Flow behind the curtain was primarily in a horizontal plane, but a significant vertical flow (i.e., "w" component) existed at the face which increased the velocity measured with the three-axis instrument.

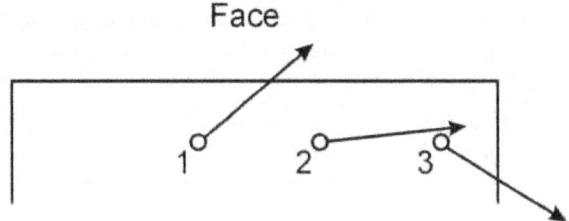

Figure 87. Flow direction at face.

Summary

This section examined instrumentation for measuring air velocities behind face ventilation curtains. Ultrasonic anemometers using one-, two-, and three-axes were compared for measurement accuracy. The effects of yaw and tilt angle were evaluated.

- Ultrasonic anemometers were adapted for use in the ventilation test gallery to measure airflow velocity and direction.
- Techniques were developed for checking instrument performance.
  - Low air velocities were measured.
  - Effects of yaw and tilt angle on flow readings were evaluated.
- Two- or three-axis anemometers are needed to accurately measure flow inby the ventilation curtain or tubing due to flow turbulence. A one-axis instrument provides accurate readings only if it is aligned with airflow.

METHANE MONITORING DURING ROOF BOLTING

Introduction

Although most frictional methane ignitions occur during the mining of coal, ignitions can also occur during roof bolting. Most of these ignitions occur when the drill bit is heated during drilling in hard or abrasive rock and comes in contact with the methane liberated from roof strata near the drilling operation. Methane concentrations must be measured with a handheld detector at least once every 20 minutes to assure that methane levels do not exceed 1%.

Laboratory Tests

Workers can remain under a supported roof if extendable probes are used to position methane detectors at the face. However, the deeper the cutting depth, the more difficult it is to reach the face with the probe. None of the techniques currently available for making these face measurements during deep cutting are easy to use or have been widely accepted. NIOSH conducted a study to examine alternative ways for monitoring methane at the face during roof bolting [Taylor et al. 1999].

NIOSH conducted tests in the ventilation test gallery to simulate airflow patterns and methane distributions with a roof bolter operating. The model mining machine was located at four locations near the face of the test gallery (Figure 88). Methane was released from the face manifold to simulate uniform face emissions (see section 1, "Test Facilities," for description of manifold) and from a hose nozzle placed against the roof near either the right or left drill booms to simulate methane release from the drill holes.

Methane concentrations were measured at 11 locations (Figure 88) which were divided into three areas:

- Face locations (1–3) were 1 ft from the roof and face.
- Sweep locations (4–7) were 1 ft from the roof and 20 ft inby the location where the bolter operators would stand (i.e., under supported roof).
- Machine locations (8–11) were 1 ft from the roof and adjacent to the T-bar. Location 11 was at the midline of the machine.

Machine locations (1 to 4), intake flows (4,000 and 7,000 ft^3/min), and curtain setback distances (28 and 40 ft) were varied to simulate a variety of operating conditions parameters.

Concentrations from all tests for a given sampling area were averaged for each release location. The results (Figure 89) show that concentrations were

- Highest at the face when gas was released from the manifold.
- Highest at the sweep location when the gas was released from the right or left drill locations.

For the manifold release tests, concentrations measured at the face were compared to concentrations measured at the sweep and machine locations. Measurements made at locations 4 (right side sweep) and 8 (right side machine) had the best correlation with the face concentrations.

During mining, most methane gas is liberated from the face. During roof bolting, generally less gas is released but most liberation occurs from both the face and the roof near the drill holes. The tests show that it is necessary to sample at the sweep and face locations to get the best estimates of gas from the face and roof.

Based on this study, an alternative sampling procedure for use during roof bolting was proposed. The procedure included use of a hand-held detector with extendable probe and a machine-mounted monitor. The detector and probe are used to make measurements no less than 16 ft inby the last area of permanently supported roof at least once every 20 minutes. The methane monitor is mounted on the roof bolter near the inby end of the automated roof support and used to continuously monitor methane levels. The final rule allowing this alternative procedure was published by MSHA in 2003.

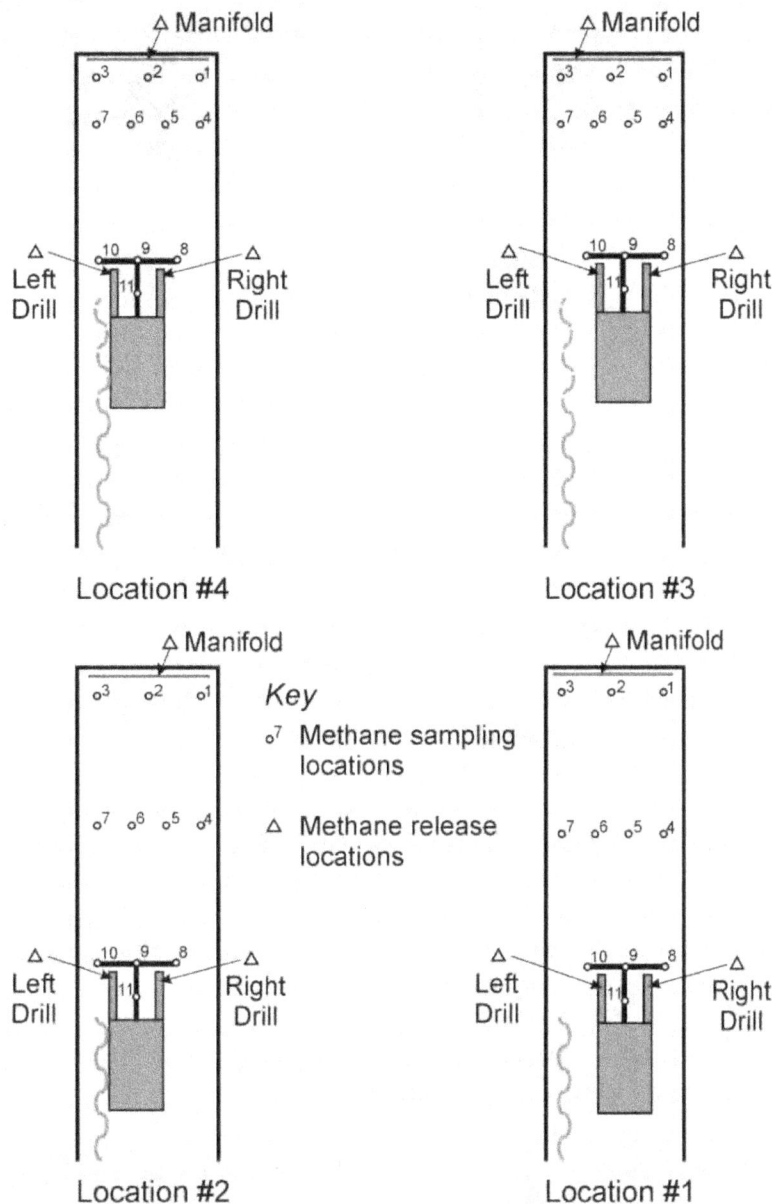

Figure 88. Methane sampling locations for roof bolter tests.

Summary

A sampling strategy for monitoring methane during roof bolting was developed and tested in the ventilation test gallery. The strategy is designed for use during bolting of deep cuts. It includes continuous monitoring on the bolting machine with a machine-mounted monitor and periodic measurements 16 ft inby the last row of bolts using a gas detector and extendable probe.

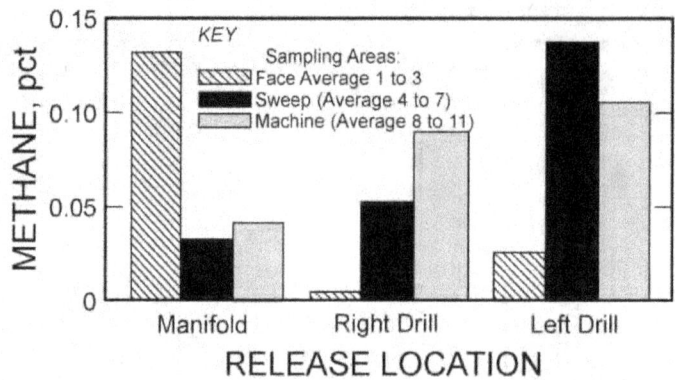

Figure 89. Effects of release location on concentration.

PRACTICAL GUIDELINES FOR THE CONTROL AND MONITORING OF METHANE GAS ON CONTINUOUS MINING OPERATIONS

Free standing fans can be used to ventilate empty headings in coal mines.

- The fan nozzle should be designed to provide maximum throw distance.
- Recirculation should be minimized by proper placement of fan inlet and/or curtains placed partway across the entry.

With blowing systems, the single most important factor on face methane dilution is the velocity of the air directed toward the face.

- For the same airflows, use of tubing rather than curtain usually provides better control of face methane, especially at longer setback distances.

With blowing and exhausting systems and with the mining machine at the face, use of scrubbers increases the amount of intake airflow reaching the mining face.

- Scrubber and spray systems should be designed to achieve efficient face ventilation for the effective removal of gas from the face.

Measurement of airflow speed and direction between the curtain and the face helps to predict methane concentrations in the face area.

- In empty entries, airflow velocity is much lower in narrower entries. More airflow should be provided during box cuts to prevent higher methane levels.

Regardless of intake flow quantity, increasing scrubber flow will reduce face methane levels if recirculation is controlled. Recirculation can be controlled by

- Minimizing leakage around the ventilation curtain.
- Directing scrubber exhaust away from the blowing curtain. With exhaust systems, the mouth of the curtain should always be outby the scrubber exhaust.

Water sprays on the mining machine should be directed to provide the best airflow across the entire face.

Methanometer response times can be measured using either of two techniques developed by NIOSH. Instruments with shorter response times more accurately measure current methane levels. The factor having the greatest effect on response times is dust cap design.

- When selecting a methanometer, the dust cap design should be examined. The cap should protect the methane sensor from dust and water but not significantly increase the response time.

Alternative methane sampling locations on the mining machine should be compared and selected based on the relative protection provided to the face workers.

Mine personnel should be provided with methane monitors that can be worn while working in areas that cannot be regularly monitored. Audible, visual, and vibration alarms for the monitors should be evaluated based on the environment in which the instruments are used.

Miners can be safely removed from a mine without exposure to excessive methane following stoppage of a main fan.

- Mines should evaluate where methane gas is likely to accumulate following stoppage of a main mine fan.

In areas between the mouth of the ventilation curtain and the face, airflow direction is constantly changing and it is difficult to accurately measure flow velocity with a single-axis anemometer (e.g., a vane anemometer).

- Following approval for underground use, multi-axis anemometers should be used to monitor airflow direction and velocity between the mouth of the ventilation curtain or tubing and the face. Multi-axis instruments should also be used to monitor flow at locations outby the mining face.

During roof bolting, if it is not practical to monitor methane levels at the mining face, methane levels should be measured with a bolter machine-mounted monitor and a detector held 16 ft inby the last row of bolts using an extensible pole.

REFERENCES

Boshkov S, Wane MT [1995]. Errors in underground air measurements. Min Eng, Vol. 47, pp. 1047–1053.

Campbell JAL [1987]. The recirculation hoax. In: Proceedings of the 3rd U.S. Mine Ventilation Symposium. University Park, PA: Pennsylvania State Univ, pp. 24–29.

Chilton JE, Taylor CD, Timko RJ [2003]. Evaluation of IYONI II methanometers. In: Proceedings of the 30th International Conference of Safety in Mines Research Institutes. Johannesburg, South Africa: South African Institute of Mining and Metallurgy, pp. 615–640.

Chilton JE, Taylor CD, Hall EE, Yantek D [2005]. Evaluation of person wearable methane monitors. In: Proceedings of the 8th International Mine Ventilation Congress. Brisbane, Australia: International Mine Ventilation Congress.

Chilton JE, Taylor CD, Hall EE, Timko RJ [2006]. Effect of water sprays on airflow movement and methane dilution at the working face. In: Mutmansky JM, Ramani RV, eds. 11th U.S./North American Mine Ventilation Symposium. Leiden, The Netherlands: Taylor & Francis/Balkema, pp. 401–406.

CFR. Code of Federal Regulations. Washington, DC: U.S. Government Printing Office, Office of the Federal Register.

Engineers International, Inc. [1983]. Testing jet fans in metal/nonmetal mines with large cross-sectional airways. By Dunn MF, Kendorski FS, Rahim MO, Mukherjee A. Engineer's International, Inc. USBM contract no. J0318015, Open File Report, No. 106-84, NTIS PB 84-196393, 132 pp.

Gillies ADS [1982]. Studies in improvements to coal face ventilation with mining machine mounted dust scrubber systems. Presented at the SME Annual Meeting, Dallas, TX, November, Preprint 82-24.

Goodman GVR, Taylor CD, Divers ED [1990]. Ventilation schemes permit deep advance. Coal, October, pp. 50–53.

Hadden JD, Hoover SM [1972]. Face ventilation survey report, Loveridge mine. Mine Enforcement and Safety Administration (MESA), 23 pp.

Hall ED, Taylor CD, Chilton JE [2007]. Using ultrasonic anemometers to evaluate face ventilation conditions. Presented at the SME Annual Meeting, Denver, Colorado, February 25–28, Preprint 07-096.

Hall EE, Timko RJ [2005]. Determining the accuracy of low velocity airflow measurements. Presented at the SME Annual Meeting, Salt Lake City, Utah, February 28–March 3, Preprint 05-80.

James RS [1959]. A continuous methane monitoring system at the working face. Min Congress J *June*:44 ff.

Kirk WS [1996]. The history of the Bureau of Mines. In: U.S. Bureau of Mines Minerals Yearbook, 1994. Washington, DC: U.S. Bureau of Mines.

Kissell FN [1979]. Improved face ventilation by spray jet systems. In: Proceedings of the Second Annual Mining Institute. Tuscaloosa, AL: University of Alabama.

Mundell RL [1977]. Blowing versus exhausting face ventilation for respirable dust control on continuous mining sections. Mine Enforcement and Safety Administration (MESA), Informational Report, No.1059.

NIOSH [1999]. Evaluating the ventilation of a 40-foot two-pass extended cut. By Thimons ED, Taylor CD, Zimmer JA. Cincinnati, OH: U.S. Department of Health and Human Services, Centers for Disease Control and Prevention, National Institute for Occupational Safety and Health, NIOSH Report of Investigations, No. 9648.

Patterson CH [1961]. Progress in ventilating continuous miner sections. Min Congress J *March*:29–32.

Taylor CD, Goodman GVR, Vincze T [1992]. Extended cut face ventilation for remotely controlled and automated mining systems. In: Khair AW, ed. New Technology in Mine Health and Safety: proceedings of the symposium held at the SME Annual Meeting, Phoenix, AZ, pp. 1–11.

Taylor CD, Rider J, Thimons ED [1996]. Changes in face methane concentrations using high capacity scrubbers with exhausting and blowing ventilation. Presented at the SME Annual Meeting, Phoenix, Arizona, March 11–14, Preprint 96-167.

Taylor CD, Rider JP, Thimons ED [1997]. Impact of unbalanced intake and scrubber flows on methane concentrations. In: Ramani RV, ed. Proceedings of the 6th International Mine Ventilation Congress, Chapter 27. Pittsburgh, PA: SME/AIME, pp. 169–172.

Taylor CD, Thimons ED, Zimmer JA [1999]. Comparison of methane concentrations at a simulated coal mine face during bolting. Mine Vent. Soc. of SA *52*(2):48–52.

Taylor CD, Thimons ED, Zimmer JA [2001]. Factors affecting the location of methanometers on mining equipment. In: Wasilewski S, ed. Proceedings of the 7th International Mine Ventilation Congress, Krakow, Poland, Chapter 97, pp. 683–687.

Taylor CD, Chilton JE, Mal T [2002a]. Evaluating performance characteristics of machine-mounted methane monitors by measuring response time. In: de Souza E, ed. Mine ventilation: proceedings of the North American/Ninth U.S. Mine Ventilation Symposium. Lisse, The Netherlands: Swets & Zeitlinger, pp. 315-321.

Taylor CD, Timko RJ, Thimons ED, Zimmer JA [2002b]. Safety concerns associated with the use of electrically powered haulage to remove workers from mines during main fan stoppages. In: de Souza E, ed. Mine ventilation: proceedings of the North American/Ninth U.S. Mine Ventilation Symposium. Lisse, The Netherlands: Swets & Zeitlinger, pp. 649–653.

Taylor CD, Timko RJ, Senk MJ, Lusin A [2003]. Measurement of airflow in a simulated underground mine environment using an ultrasonic anemometer. Presented at the SME Annual Meeting, Cincinnati, OH, February 24–26, Preprint 03-065.

Taylor CD, Timko RJ, Thimons ED, Mal T [2005]. Using ultrasonic anemometers to evaluate factors affecting face ventilation effectiveness. Presented at the SME Annual Meeting, Salt Lake City Utah, February 28–March 3, Preprint 05-80.

Taylor CD, Chilton JE, Hall E, Timko RJ [2006]. Effect of scrubber operation on airflow and methane patterns at the mining face. In: Mutmansky JM, Ramani RV, eds. 11th

U.S./North American Mine Ventilation Symposium. Leiden, The Netherlands: Taylor & Francis/Balkema, pp. 393–399.

Taylor CD, Chilton JE, Martikainen AL [2008]. Use of infrared sensors for monitoring methane in underground mines. In: Wallace KG Jr., ed. Proceedings of the 12th U.S./North American Mine Ventilation Symposium. Reno NV: University of Nevada, pp. 307–312.

USBM [1938]. Some observations on coal-mine fans and coal-mine ventilation. By Harrington D, Denny EH. Washington, DC: U.S. Bureau of Mines, Information Circular, No. 7032.

USBM [1940]. Some information for miners about coal-mine ventilation. By Miller AU. Washington, DC: U.S. Bureau of Mines, Information Circular, No. 7137.

USBM [1950]. Achievements in mine safety research and problems yet to be solved. By Fieldner AC. Washington, DC: U.S. Bureau of Mines, Information Circular, No. 7573.

USBM [1958]. Auxiliary ventilation of continuous miner places. By Stahl RW. Washington, DC: U.S. Bureau of Mines, Report of Investigations, No. 5414.

USBM [1966]. Face ventilation in underground bituminous coal mines, performance characteristics of common jute line brattice. By Dalzell RW. Washington, DC: U.S. Bureau of Mines, Report of Investigations, No. 6725, pp. 29.

USBM [1968]. Face ventilation in underground bituminous coal mines, airflow characteristics of flexible spiral-reinforced ventilation tubing. By Peluso RG. Washington, DC: U.S. Bureau of Mines Report of Investigations, No. 7085.

USBM [1969a]. Face ventilation in underground bituminous coal mines. By Luxner JV. Washington, DC: U.S. Bureau of Mines, Report of Investigations, No. 7223.

USBM [1969b]. Studies on the control of respirable coal mine dust by ventilation. By Kingery DS et al. Washington, DC: U.S. Bureau of Mines, Technical Progress Report, No. 19.

USBM [1973]. Evaluation of a machine-mounted dust collector. By Tomb TF, Martonik JF, Taylor CD. Washington, DC: U.S. Bureau of Mines, Report of Investigations, No. 7788.

USBM [1978]. Jet fan effectiveness as measured with SF6 tracer gas. By Matta JE, Thimons ED, Kissell FN. Washington, DC: U.S. Bureau of Mines, Report of Investigations, No. 8310.

USBM [1987]. Deep cut mine face ventilation. By Ruggieri SK et al. Washington, DC: U.S. Bureau of Mines, Final Contract Report No. HO348039.

USBM [1992]. Jet fan ventilation in very deep cuts—a preliminary analysis. By Goodman GVR, Taylor CD, Thimons ED. Washington, DC: U.S. Bureau of Mines, Report of Investigations, No. 9399.

Volkwein JC, Thimons ED [1986]. Extended advance of continuous miner successfully ventilated with a scrubber in a blowing section. Presented at the SME Annual Meeting, St Louis

Missouri, Preprint 86-308.

*Delivering on the Nation's promise:
safety and health at work for all people
through research and prevention*

To receive NIOSH documents or more information about
occupational safety and health topics, contact NIOSH at

1 800 CDC–INFO (1–800–232–4636)
TTY: 1–888–232–6348
e-mail: cdcinfo@cdc.gov

or visit the NIOSH Web site at **www.cdc.gov/niosh**.

For a monthly update on news at NIOSH, subscribe to
NIOSH *eNews* by visiting **www.cdc.gov/niosh/eNews**.

DHHS (NIOSH) Publication No. 2010-141

SAFER • HEALTHIER • PEOPLE™

www.ingramcontent.com/pod-product-compliance
Lightning Source LLC
Chambersburg PA
CBHW081839170526
45167CB00007B/2840